W9-CFO-526

Taking Chances

Taking Chances

Winning with
Probability

JOHN HAIGH

Reader in Mathematics and Statistics,
The University of Sussex

OXFORD

UNIVERSITY PRESS

Great Clarendon Street, Oxford OX2 6DP

Oxford University Press is a department of the University of Oxford
and furthers the University's aim of excellence in research, scholarship,
and education by publishing worldwide in

Oxford New York

Athens Auckland Bangkok Bogotá Buenos Aires Calcutta
Cape Town Chennai Dar es Salaam Delhi Florence Hong Kong Istanbul
Karachi Kuala Lumpur Madrid Melbourne Mexico City Mumbai
Nairobi Paris São Paulo Singapore Taipei Tokyo Toronto Warsaw

and associated companies in Berlin Ibadan

Oxford is a registered trade mark of Oxford University Press

Published in the United States
by Oxford University Press Inc., New York

© John Haigh, 1999
First published 1999
Reprinted in paperback (TSP) 1999
Reprinted as OPB 2000

All rights reserved. No part of this publication may be reproduced,
stored in a retrieval system, or transmitted, in any form or by any means,
without the prior permission in writing of Oxford University Press.
Within the UK, exceptions are allowed in respect of any fair dealing for the
purpose of research or private study, or criticism or review, as permitted
under the Copyright, Designs and Patents Act, 1988, or in the case
of reprographic reproduction in accordance with the terms of licenses
issued by the Copyright Licensing Agency. Enquiries concerning
reproduction outside those terms and in other countries should be
sent to the Rights Department, Oxford University Press,
at the address above.

This book is sold subject to the condition that it shall not, by way
of trade or otherwise, be lent, re-sold, hired out, or otherwise circulated
without the publisher's prior consent in any form of binding or cover
other than that in which it is published and without a similar condition
including this condition being imposed on the subsequent purchaser

A catalogue record for this book is available from the British Library

Library of Congress Cataloging in Publication Data
Haigh, John, Dr.
Taking chances: winning with probability/John
Haigh
Includes bibliographical references and index.
1. Probabilities–Popular works. I. Title
QA273. 15. H35 1999 519.2–dc21 98–39018

ISBN 0 19 850292 3 (Hbk)
ISBN 0 19 850430 6 (TSP Pbk)
ISBN 0 19 850291 5 (OPB)

Printed in Great Britain by Biddles Ltd, Guildford and King's Lynn

'... misunderstanding of probability may be the greatest of all impediments to scientific literacy.'

(Stephen Jay Gould)

Preface

You apply probability to almost every conscious decision you make. What you wear will depend on your assessment of the weather outlook. When you cross the road, you ensure that the chance of being struck by a vehicle is acceptably small. Your spare light bulbs are there 'just in case'. The whole of the insurance industry, be it property, possessions, life or whatever is based on judgements about the balance of risks. If humans were not naturally so good at working intuitively with many ideas of probability, modern civilization would never have evolved.

But despite their frequently good intuition, many people go wrong in two places in particular. The first is in appreciating the real differences in magnitude that arise with *rare* events. If a chance is expressed as 'one in a thousand' or 'one in a million', the only message registered may be that the chance is remote—and yet one figure is a thousand times bigger than the other. Another area is in using partial information: you might be asked to guess, from looking at a photograph, whether a striking young woman is a model or an office worker. It is tempting to go for the former, forgetting that there are far more office workers than models to choose from.

Anyone who buys Lottery tickets, plays games with cards or dice, or bets on the Grand National, is aware of the operation of random chance. Even if we do not sit down to make the calculations, we have some intuition and experience that tells us what is possible, what is likely, and what is typical. These assessments are usually good enough for practical purposes, but without a sound approach to combining and interpreting probabilities, it is easy to blunder. Suppose you are a member of a jury, where the defendant is accused of a serious crime such as rape or murder. The evidence is said to be such that the chance of it arising if the defendant were innocent is one in a million. Some people might believe that this is equivalent to saying that the chance the defendant is innocent, given this evidence, is one in a million. Nonsense. That piece of logic is akin to equating 'The chance that Jean forgets her coat when it rains is 1%' with 'The chance that it rains when she forgets her coat is 1%'.

In this book, I describe a number of common experiences in which chance plays a role. My aim is to entertain, enlighten and inform (not necessarily in that order). Some examples will be familiar, others less so, but all have some connection with probability. Each chapter has some common theme, sometimes ending with a postscript, a diversion into some probabilistic corner. Often, an idea can be communicated using some game between two or more individuals or teams. The notion that you can gain an edge in such games by being good at working with probability usually appeals to our competitive instincts.

This book is aimed at the 'layman', interpreted as someone untrained in science or mathematics, but wishing to reason probabilistically with confidence. It is said that every mathematical expression in a 'popular' book halves the readership. I could not achieve my objectives by eschewing all such expressions, but I have sought to keep symbols to the minimum in the main text. There are five appendices into which I have shamelessly drawn upon more mathematical familiarity. Not every reader will run at the sight of an x. In a number of places within the text some material is separated out into a box. These boxes are usually more mathematical than their surroundings, and can always be ignored without cost to the main story. I hope many readers will not be inhibited, and will seek to come to grips with what is written there. Some powerful ideas are surprisingly accessible, if you try.

In the interests of ease of reading, I have rounded various numbers. If a probability is said to be 50%, that might either be exact (e.g. the chance of a fair coin showing Heads) or approximate (e.g. the chance a soccer match produces a home win). The context will usually make it clear which alternative applies. I have tried to make the approximations accurate to the number of significant figures given: 77% means between 76.5% and 77.5%, while 60% could represent the range from 55% to 65%. When manipulating numbers, best practice is to work exactly for as long as possible, and then round at the final stage. I have also adopted the common usage of 'K' to denote 1000 when discussing sums of money, so that a prize of £25 000 may be written as £25K.

Look on the whole book as a challenge to solve the problems stated, before the text does that for you. Some chapters end with a few problems for you to test yourself on, to judge whether you have assimilated the ideas and main conclusions. Solutions are given later. You ought to read Chapter 1 first, Chapter 6 before Chapter 10, Chapter 8 before Chapter 12, and Chapter 13 last; otherwise, the order is largely unimportant. This is not a textbook in which a series of ideas are systematically introduced, reaching a grand and deep finale.

I could not list all those whose writings, lectures or talks have helped me understand the ways in which probabilists see the world. This book has no claims to originality, except anywhere I specifically make one. The list of references contains only those works definitely used in the text. But it is a pleasure to acknowledge the direct help of so many people. Susan Harrison, of OUP, encouraged me to start this book and to see it through. I also thank Geoff Adlam, John Barrett, Trevor Beaumont, Mike Birkett, Toby Brereton, John Croucher, Frank Duckworth, John Gwynn, Bob Henery, Leighton Vaughan Williams, Harry Joe, Robert Matthews, Hugh Morton, Louise O'Connor, Hans Riedwyl, Jonathan Simon, Ian Walker, Greg Wood, and Bill Ziemba for responding to my requests for information, or supplying it anyway. Richard Lloyd performs a wonderful service by constantly updating his internet site devoted to the National Lottery—see http://lottery.merseyworld.com/. My formal teachers—Bert Lodge, Jeffrey Lloyd, Geoffrey Whitehouse, Ken Senior, Phyllis Bowman, John Lewis, David Kendall, and John Kingman—got me interested in mathematics, showed me its power, and led me in probabilistic ways of thought. My greatest mathematical debt is to someone I never met: William Feller, whose books inspired a generation of probabilists. Adam and Daniel Haigh taught me how to use the family computer to produce the typescript. And if any chapters have hit the right level, it is because of the tactful comments of Kay Haigh, who acted as guinea pig to my first offerings.

Despite my best endeavours, this book will not be error-free. I hope that none of the mistakes are misleading. But since probability can be a subtle subject, if I have made some slips of principle, do not hesitate to tell me about them. Greater minds than mine have erred in this field.

Brighton   J.H.
September 1998

Note (November 1999)

This (softback) edition is unchanged, except for minor corrections and some updating. A list of updated corrections will be maintained at http://www.maths.susx.ac.uk/Staff/JH/.

Contents

CHAPTER 1

What is probability?

Whatever it is, you cannot ignore it. The insurance premium for your house, car, or life reflects the probability the insurers will have to pay out, and how much. Should you have a vaccination against winter flu? You must balance the risks of side-effects or possible reaction to the vaccination against the consequences of declining the protection. Members of a criminal jury are asked to convict only if they believe guilt is established 'beyond reasonable doubt'. In a civil case, the relevant criterion may be 'on the balance of probabilities'. You buy, or do not buy, tickets for the National Lottery partly on impulse or for fun, but also partly on your assessment, however vague, of your chance of winning a substantial sum. In card games such as poker or bridge, you expect to play better if you can make a realistic judgement of the likelihood that an opponent holds certain cards. Many decision problems, serious or frivolous, are better informed by a sound understanding of the notion of probability. My credo is that knowledge is preferred to ignorance, and that one way of achieving good familiarity of probability is through exposure to a variety of 'games' in which chance plays a role. I aim to illustrate ways in which your winning chances can be assessed, and sometimes increased.

Merely being an expert on probability is no guarantee that you will make the right decision. Sometimes, all you gain from knowledge is a better appreciation of why you were wrong. So be it. But, on balance, your decision-making in life as well as in games ought to improve if you can make good assessments of the chances of the various outcomes. This book is not an academic study of probability theory, rather a series of challenges whose solutions will involve probabilistic arguments.

English thrives from the ability to say the same thing in different ways. You might shuffle a pack of cards, select the top one, and assert: '*the probability that it is a Spade is one-quarter*'.
Each of the following has precisely the same meaning as this:

- The chance that it is a Spade is one-quarter.

- The odds that it is a Spade are three to one against.
- A non-Spade is three times as likely as a Spade.

More elliptical ways of saying the same thing might use the words 'risk' or 'likelihood', and a thesaurus would send you on a word-chase including 'prospect', 'possibility', and so on. But however we formulate that statement, what do we mean? And what might lead us to choose one-quarter, rather than any other value?

I would choose 'one-quarter' for this probability through an idealized model. My model has 52 identically composed cards, 13 of them Spades, and all possible arrangements of the shuffled deck are taken to be equally likely. If all that were true, then one-quarter of these equally likely arrangements have a Spade on top—hence my choice. I know my model cannot be exactly correct in all respects, but I hope it is near enough to reality not to give a misleading answer. A perfect model of the physical world was not required to land objects on the Moon and Mars.

This experiment with cards can be repeated as often as we like. In such a case, we can test the validity of a probability statement by looking at a long sequence of experiments. For an event to have probability one-quarter, we expect it to arise one-quarter of the time, on average. Emphatically, we do *not* mean it must occur exactly once in every block of four repetitions: it might occur several times in succession, it might not occur over a dozen experiments. A problem with this approach is to know what those weasel words 'on average' or 'long sequence' mean. Are 100 experiments enough? 10 000? Unfortunately, there is no simple way to say how long is long enough.

We also like to speak of the probability of one-off events, such as whether the index of Stock Exchange prices will rise by at least 10% over the next calendar year, or whether Brazil will win the next World Cup at soccer. The idea of long-run average frequency is irrelevant when there is no way of reconstructing the current circumstances. Nor are we restricted to discussing what might happen in the future: a statement such as 'The probability that Shakespeare wrote *Macbeth* is 80%' is quite meaningful to express the opinion of a scholar.

To think of 'probability' as describing a *degree of belief* reconciles all these approaches. The stronger the degree of belief in an event, the larger the probability associated with it. If I wish to assess my own degree of belief that the Stock Exchange will rise by a given amount, I can ask a bookmaker to give me odds. Perhaps I am offered odds of two to one. In other words, if I bet £1 and win, I will receive £3 (my original £1, plus £2 for the two to one odds). My reaction to the offer tells me about the probability I attach to the event in question. If I feel that making a bet is favourable at these odds, I am attaching a probability of more than one in three; if I shy away, thinking that the bet is poor value, then my degree of belief is less than one in three. While I am considering the attractiveness of the bet, you could be doing the same, and your opinion could easily differ from mine. Probability is personal. In the example about selecting a Spade from a pack of cards, our personal probabilities are likely to coincide at one in four, because we have more or less identical models for the experiment. In other circumstances—especially when we have different information—our assessments of the probabilities may differ greatly. The trainer of a racehorse will have a different opinion of his horse's chances from either the bookmaker or the average punter.

In examples concerning dice, coins, cards, etc., there will be a wide-spread consensus about the appropriate model, and so little disagreement about the probabilities derived. People may have different reasons that lead them to conclude that a particular roulette wheel is fair, but those who are prepared to agree that all 37 numbers are equally likely will analyse all permitted bets using the same model. This position will be the case for most of this book. But in some applications, experts will have honest differences in their judgements. Discovering whether these differences have a serious effect on any decisions we make, or conclusions we draw, is a vital part of an analysis.

Think straight

The Victorian scholar Francis Galton offered a warning example against loose thinking. If you toss three fair coins, what is the probability they all show the same, either all Heads or all Tails? Consider the following sloppy argument:

> At least two of the three coins must be alike, either both Heads or both Tails. The third coin is equally likely to fall Heads or Tails, so half the time it is the same as the other two, and half the time it is different. Hence the probability all three are the same is one-half.

Pinning down the error in this flawed argument requires a logical approach. One way is to label the coins as Red, Blue and Green, and then list the possible outcomes with the coins considered in that order. By distinguishing the coins, you avoid the mistake that Galton provocatively offered, and you will see that there are eight outcomes, {HHH, HHT, HTH, THH, HTT, THT, TTH, TTT}, of which just two are all alike. That leads to a different answer. The chance that all three coins fall alike is two in eight, or equivalently, one-quarter.

The flaw in the sloppy argument is that loose phrase 'the third coin' at the beginning of the second sentence. If we do not distinguish the coins from each other, how do we know which one is this third coin? If exactly two coins show Heads, it is plain that the other one must be the third coin—but it follows also that this coin shows Tails; it is not 'equally likely to be Heads or Tails'. And if all three coins show Heads, whichever is labelled as the third coin must also show Heads. It too is not equally likely to show Heads and Tails.

Such loose thinking can cost money. Suppose Andy reckons the chance is one in four, and is (rather ungenerously) willing to pay out at odds of two to one if all three show the same. He expects to win money on this bet: if £1 is staked, he anticipates keeping the money in three games out of four, and making a £2 loss the other game. On average, four games will leave him £1 in profit. If Brian believes the fallacious argument, he will be more than happy to accept this bet. He feels that half the time he will lose £1, but half the time he will win £2, so the bet favours *him*. Both are willing participants, but Brian's analysis is flawed. The longer he persists in playing, the more his losses will increase. He should review his thought processes.

The inviolable rule

The vocabulary of probability includes descriptive words and phrases such as 'unlikely', 'possible', 'probable', 'almost certain', etc. But what you mean by 'very likely' need not be the same from day to day, and will also differ from what I mean. To understand each other, we use numbers.

Probabilities are measured on the scale of zero to one. The words 'impossible' and 'probability zero' mean the same thing. I attach probability zero to the event that someone could travel backwards in time to meet Mozart; you are entitled to your own opinion. At the other end of the scale, 'probability one' is equivalent to 'certain'. I regard it as certain that Elvis Presley is dead; some people appear willing to wager money that he will appear alive within ten years (doubtless on water skis, towed across Loch Ness by the monster). Whatever your beliefs about real or hypo-thetical events, no probability outside the range from zero to one makes any sense. This is an inviolable rule.

(Nevertheless, I have a personal embarrassment here. As a neophyte university lecturer, I once offered a 'clever' examination question about a relationship between two probabilities. I could solve it beautifully, and the answers looked sensible. Unfortunately, a consequence of my solution was that two *other* probabilities violated this rule! My bacon was saved by a more experienced colleague, whose checks ensured my question got no further than his desk.)

You can express probabilities as odds, fractions, decimals, or percent-ages in the range from 0% to 100%, according to taste. Fractions arise naturally (Appendix II) when all outcomes are assumed equally likely, such as in throwing dice, dealing cards, or spinning roulette wheels. In these problems, the calculations are usually easiest when the probabilities are left as fractions; it is nearly always a mistake to turn them into decimals at an early stage. But if you wish to compare probabilities, percentages, decimals, or fractions with a common denominator can all be useful.

Some working methods

Appendix II gives a fuller description, and some examples. But you can get a long way with two central ideas. The first is to do with *exclusive* events: if it is impossible for two things to happen simultaneously, they are exclusive. For example, if you draw one card from a pack, it could be either a Heart or a Diamond, but not both—those events are exclusive. It could be a Club, or a King—and those are not exclusive, as the card could be the King of Clubs. When events are exclusive, to get the chance that either will occur, you just add together the separate chances. If the chance of a Heart is one-quarter, and also the chance of a Diamond is one-quarter, the chance of a red card adds up to one-half.

The second central idea is *independence*. Two events are independent if knowing whether or not one of them occurs has no effect on the probability of the other. For example, if I toss a coin, and you toss a coin, the outcomes can reasonably be taken as independent. But if I draw a card from a pack, and you draw a second card from those remaining, these outcomes will not be independent: the knowledge that my card is, say, an

Ace reduces the proportion of Aces in what is left, so your chance of an Ace is affected. It is usually obvious when two events can, or cannot, reasonably be taken as independent. The working definition of two events being independent is that *the probability that both occur is the product of their individual probabilities.*

An ordinary die shows a Six with probability 1/6. Roll it twice, independently, and the probability both are Sixes is $(1/6) \times (1/6) = 1/36$. Suppose there is a major oil spillage at sea about one year in three and a London team wins the FA Cup about one year in four. It is hard to imagine how these might affect each other, so it seems reasonable to suppose these events are independent. Hence, for a year chosen at random, the probability there is both a major oil spillage and a London team winning the FA Cup is about $(1/3) \times (1/4) = 1/12$.

Occasionally, it is a surprise to find that some events are independent. Roll an ordinary fair die, and consider the two events:

- Get an even number, i.e. 2, 4, or 6.
- Get a multiple of three, i.e. 3 or 6.

Even on the same roll, these turn out to be independent! This means that to be told the outcome is even makes no difference to the chance it is a multiple of three, and vice versa. With this die, each of the six separate faces has probability 1/6; as there are three even numbers, the total chance of an even number (adding the three exclusive chances) is $3/6 = 1/2$. In the same way, the chance the outcome is a multiple of three is $2/6 = 1/3$, and these two probabilities multiplied together give 1/6. But the only way *both* events can occur is to roll a Six, which also has probability 1/6. So these events are indeed independent, as the chance they both occur is the product of the individual chances.

Good games shops also stock dice with different numbers of faces. Replace the ordinary six-sided die by a tetrahedron that gives equal probability to each of the four sides, labelled $\{1, 2, 3, 4\}$. This time, the same two events are *not* independent. If you know the outcome is even, it then becomes impossible for it to be a multiple of three (and also vice versa). Formally, using the above definition of independence, the chance of an even number is still 1/2 (two even numbers among the four available), but the chance of getting a multiple of three changes to 1/4, since only one of the four outcomes is a multiple of three. Multiplying these probabilities together gives the answer $(1/2) \times (1/4) = 1/8$. But on this four-sided die, it is not possible to have an outcome that is both even and a multiple of three, since the die has no multiple of six. The chance both events occur is zero—which differs from 1/8.

Use a diamond-shaped die with eight sides, and the two events become independent again! This time, even numbers will occur half the time, and multiples of three occur one time in four. To have both an even number, and a multiple of three, we need a multiple of six. On this eight-sided die, there is just one such multiple, 6 itself, so a multiple of six has chance one in eight. And since $(1/2) \times (1/4) = 1/8$, the events are independent, according to the working definition.

With this octahedral die, the information that the outcome is a multiple of three means that only 3 or 6 can occur. And then the outcome will be the even number, 6, half the time. The chance of an even number is one-half, whether or not the outcome is a multiple of three. In the same way, if we are told that the outcome is even, there are four possibilities {2, 4, 6, 8} which contain just one multiple of three. So being told the outcome is even makes no difference to the chance of getting a multiple of three; it is 1/4 with or without that information.

This series of examples shows two events—getting an even number, or getting a multiple of three—that might or might not be independent, according to the number of faces on the die. Take care to specify a model completely, when working with probabilities. To decide intuitively whether two events are independent, ask yourself one question: if you know for sure that one of them has occurred, does that alter the chance of the other in any way? If the answer is 'No', the events are independent.

Long runs

Although I have stressed that probability simply expresses your own degree of belief, it is useful to know what to expect in a long sequence of repetitions, such as in tossing a fair coin. Here, the probability of Heads on any toss will be 1/2, but what do we expect to see over thousands or even millions of tosses? Will the proportion of Heads settle down to one-half? The short answer is 'Yes', but that needs some elaboration.

Demanding exactly 500 000 H and 500 000 T in one million tosses is rather unreasonable. But if you ask for a fixed *percentage* on either side of the middle, the picture changes. Suppose you want the number of Heads to be between 49% and 51% of the total. In an experiment with just 100 tosses, this is asking for either 49, 50, or 51 Heads. If you repeated this experiment a large number of times, you would find the proportion of Heads fell in this narrow range about 24% of the time.

Increase the size of the experiment to 1000 tosses. Then anything from 490 to 510 is acceptable, and this will happen about 50% of the time. With 10 000 tosses, the permitted range is 4900 to 5100, and now the success rate is over 95%. With a million tosses, the Romanovs will regain absolute power in Russia before the proportion of Heads falls outside the 49% to 51% range.

The same sort of thing holds with even more stringent proportions: if you look for the proportion of Heads to fall between 49.9% and 50.1%, you are asking a lot with 1000 tosses (only 499 or 500 or 501 Heads fit the bill), but you would be astonished not to meet the requirement with ten million tosses. The proportion of Heads will, eventually, be as close to 1/2 as you wish. What you cannot expect is exact equality of Heads and Tails, or for the number of Heads always to be within any fixed number, say 20, of its average, when a coin is tossed millions of times. In fact, the reverse holds; if you toss a fair coin often enough, the absolute difference in the numbers of Heads and Tails will eventually exceed any number you care to name. It is the *proportion* of Heads that stabilizes.

It is the same with probabilities other than one-half. Drawing a single card from a well-shuffled pack should lead to the chance of a Spade being one-quarter, or 25%. If you repeat this experiment a large number of times, how often do you expect between 24% and 26% of them to result in a Spade? As with coins, that depends on how many times you select a card. Doing the experiment only ten times, you could *never* get a Spade between 24% and 26% of the time (that would require between 2.4 and 2.6 Spades!). Do it 100 times, and the permitted range of 24, 25, or 26 Spades will be met about one time in four. With a million experiments, the range is so wide—240 000 to 260 000—that the chance of falling outside it is comparable to the chance of a Shetland pony winning the Grand National.

To summarize: suppose an experiment is capable of being repeated independently and under identical conditions as often as we like, and some event of interest has a particular probability, x. Repeat this experiment, keeping track of the running proportion of times the event occurs. This proportion might oscillate widely to begin with, but we are interested in the long-run behaviour. Take any fixed interval, however small, enclosing the probability x, and note whether the running proportion falls inside the interval or outside it. Whatever the initial behaviour, there will come a stage when not only does the proportion fall inside this interval, it *remains* inside it thereafter. If there is such a thing as the law of averages, I have just described it. It only refers to what will happen in the distant future, and carries no inferences about the short term. The stock phrase 'by the law of averages' is bandied about in conversation, usually quite inaccurately. It is invoked to assert that because Heads and Tails are expected to be equal in the long run, they had better start getting more equal next toss! False. This 'law' says nothing about the next toss, or even the next hundred tosses. Frequencies of Heads and Tails will settle down to their average, but only 'eventually'.

If we can repeat an experiment in this fashion as often as we like, this knowledge of long-run behaviour enables us to estimate probabilities. We keep track of the proportion of times an event occurs over as many repetitions as we can afford. This proportion can be taken as an estimate of the underlying probability. Sometimes this estimate will be too big, other times too small, on average it will be just right (rather like Goldilocks sampling the Three Bears' porridge). But the more repetitions we make, the better we can expect the estimate to be.

Some experiments can be tried once only. A weather forecaster may assert that the chance of rain tomorrow is 50%. She will have reached that conclusion from studying weather maps, satellite data, previous weather patterns at the time of the year, and so on. She has some model, which leads to her prediction. She cannot test her assertion in the same way as she could test her belief about a coin falling Heads. 'Tomorrow' is one specific day, and either it rains or it does not. She can never know whether her assessment was accurate. But she might keep a cumulative record of all her predictions over several years. Maybe there are 100 occasions when she thought rain has a 50% chance; on another 80 days, she thinks the chance of rain is 25%, and so on. Matching the actual weather against her predictions, grouped together in this manner, is not very different from repeating the same experiment a number of times—providing her ability at these forecasts remains fairly consistent. Although any particular claim about the chance of rain cannot be investigated in exactly the same way as a claim about chances in repeatable experiments, we can still make judgements about the collective accuracy of her claims.

Averages and variability

Over much of science, the most useful ideas are the simplest, and this surely holds in probability, the parent of statistics. When playing a game, you might know all the possible amounts you could win, and what their chances are, but the brutal truth of whether the game favours you or your opponent usually comes from looking at an *average*. For many decisions in the field of statistics, an average will be more use than any other single quantity. I take it for granted that you have some intuitive notion about averages, from your experience of life: you will know the average time of your frequent journeys, your average supermarket bill, the average frequency you win at various games. This intuitive notion will suffice for most purposes, but sometimes we need to be precise, so then Appendix III gives formal ways to compute an average.

The draw for the fifth round of the FA Cup is about to be made. There are 16 teams, leading to eight matches. Your task is to pair the teams off, in an attempt to guess as many as possible of the actual matches in the real Cup draw. You are not asked which teams will be drawn at home, just which pairs will be selected. I am prepared to pay you £1.50 for each correct guess; how much would you be prepared to offer for the right to make the eight guesses? Write down your bid, before you read on.

The most you can win is £12, when all your guesses are correct, but you might win nothing at all. If you seek to work out the different amounts you might win, with their respective chances, you may well find the task rather daunting. But this detail is not necessary. Aside from the pleasure of participating, what will dominate your thoughts is how much you should win, on average. If your entrance fee is less than this average, the game favours you, but if it is more, the game favours me.

If you are pretty good at averages, you will have lighted on 80p as your average winnings, so perhaps you are willing to pay anything less than that. Appendix III shows where this answer of 80p comes from, but the central point here is that this answer cuts through all the detail of the many alternative outcomes, and gives a simple decision criterion. You will be willing to pay any amount up to 80p; I am happy to accept at least 80p. That sum leads to a fair game, in the sense that neither of us expects to win money from the other.

Change this game a little. Instead of offering a reward for each correct guess, there is a *threshold*. If you guess below the threshold, you win zero, but three possible games are:

(1) For at least two correct guesses, you win £7.94.

(2) For at least four correct guesses, you win £344.66.

(3) For guessing all eight, you win £1 621 620.

(Stretch a point, and imagine I have the capacity to pay such rewards.)

These rewards have been chosen to make your average win still 80p, so that would be a fair entrance fee. The *average* result is the same, in these three games and the original one, but all four games will not be equally attractive to you. One difference is in their *variability*.

In the original game, you will win something just over 40% of the time; your winning chances in these new games are about 10%, 2% and one in two million. A very cautious person is likely to prefer the original game; you won't go many games without winning something. The less cautious you are, the more attractive is game (3). Perhaps you do not mind a dozen losses of 80p, in exchange for a dozen prospects, albeit remote, of winning such a large sum.

Game (3), turned on its head, has excellent analogies with insurance. If you drive a car, you may cause damage to property or other people ('third parties') that would cost vast sums in compensation. Most of us have no prospect of paying huge amounts, so we pay a premium to an insurance company. This premium typically covers many other risks—damage to the car itself, lesser amounts of compensation—but focus on the specific point that we might cause damage costing £1 million to rectify. An insurance company may estimate that chance as 1 in 100 000 over a year, and fix an annual premium of £20.

Set this up as a game. You pay £20, and most of the time you win zero. There is a chance, 1 in 100 000, that you 'win' £1 million from the insurance company to pay for the damage you have caused. So the average amount you win is £10. In cash terms, the game is unfair, but the more cautious you are, the more you accept it is in your interests to play it! It is not the *average* payout that concerns you, it is the possibility of having to find £1 million that makes paying £20 the safe option. Playing this game eases your mind: you are protected from disaster.

The insurance company is also content, if it has enough customers. With a base of ten million customers, each paying £20, it receives £200 million. On average, since each customer carries a risk of 1 in 100 000, just 100 customers will claim on their insurance, and this costs £100 million. The company expects to be well in profit. But if over 200 customers make claims, the company faces a loss. How likely is this to happen? The company's actuary knows that, on these figures, there is no danger of disaster. With an average of 100 claims, he would not be surprised to see 110 or 115, but a year with even 150 or more claims will arise less than once in a millennium.

The arithmetic for a smaller company, say one hundredth as large, is different. Its annual premium income is £2 million and on average one claim of £1 million is made. This is exactly in proportion to the larger company, but now three or more claims will be made about one year in 12. In such a year, payouts exceed premiums. That is uncomfortably frequent, and has the potential for disaster for both parties. The variability in possible payouts for this small company is too great. Insurance companies increase their customer base to reduce their chance of exposure to bankruptcy.

This is an important general point. If you play a game once only, the average outcome is useful to know, but it will be well short of the full story. Over one game, or just a few games, the variability of the outcomes can be much more important. But over a large number of games—the insurance company is playing a separate game against each of its customers—the average dominates. The more variable the outcomes, the more games are needed before the average becomes king.

CHAPTER 2

Lotteries

Mathematicians become accustomed to requests for good strategies to play the Lottery. One main hope is that statistics can be used to predict which numbers will be drawn in the future. The autobiography of Len Shackleton, one of my boyhood heroes, includes a celebrated chapter headed 'What the Average Director Knows about Football': the chapter was simply a blank page. Any chapter from me entitled 'How to win more often' would also be a blank page, apart from the single line 'Buy more tickets'. But there are ways in which a mathematical approach to the Lottery can help gamblers, as I shall show.

The UK National Lottery began in November 1994. It has been a huge commercial success, with over 90% of the adult population having bought at least one ticket, and regular sales (September 1998) of over £80 million each week. Its logo is widely recognized, and the seductive appeal 'It could be you' reminds us how easy it is to buy a chance to win a jackpot prize. Several hundred new millionaires owe their status to Lottery wins—the top prize won by a single ticket is over £22 million.

The game follows the format of similar ones around the world. For completeness' sake, I will describe the UK rules. To play in a particular draw, you pay £1 to select six different numbers from the numbers 1 to 49. Call this is a 6/49 Lotto game. For each draw, a machine selects six different balls from a set also numbered 1 to 49. You win a prize if at least three of your numbers match those chosen by the machine. The more numbers you match, the larger your prize should be. If you match all six, you stand an excellent chance of winning a very substantial sum.

There is a second way you might win an amount large enough to change your lifestyle. The machine selects a seventh 'Bonus' number, and if you have chosen that number, along with five of the first six drawn, you will win the second prize. Otherwise, the Bonus number is irrelevant.

The organizers go to great lengths to try to demonstrate that numbers are selected completely at random, and with no opportunity for deceit. So far as is possible, the balls are identically composed, of equal size and weight, dry, and with hard, smooth surfaces. There are eight sets of these numbered rubber balls, and four different machines, simple transparent plastic tubs, to mix them thoroughly. Only on the day of the draw are the particular machine and the set of balls chosen. The draw itself is shown live on television, witnessed by an audience from the general public. Afterwards, the machine is emptied in the presence of an independent auditor who certifies that the complete set of balls has been replaced in a sealed case. These balls are tested by the Weights and Measures Department before being offered for use again. Between draws, all the sets of apparatus are kept in sealed containers.

Probability and statistics can help in the three questions:

- What are your winning chances?
- If you play, what tactics might increase your winnings?
- How can we tell whether the numbers selected by the machine are really random?

Your chance of winning

The appeal of the Lottery is the size of the jackpot prize. For many players, the sole fact of any importance is that huge prizes have been paid out. But any rational analysis has to take account of the chance of winning, as well as what might be won.

As you see the balls rotate in the machine, it is easy to accept that each of them has the same chance of being chosen first. Perhaps number 16 is chosen. Then each of the remaining 48 balls are equally likely, and so on. When six balls have tumbled out, every collection of six numbers should have the same chance of selection. This means that {1 2 3 4 5 6} has exactly the same chance as the apparently more random {3 4 20 38 48 49}. If you doubt this (and you would not be alone), then you believe that some combinations are inherently more likely than others. Perhaps you should list some combinations that have a below-average chance, and justify your choices.

I find this appeal to the symmetry of the set-up convincing. I see no good reason to pick on one combination as being more or less likely than any other. However, I know from long experience that if someone is convinced that {1 2 3 4 5 6} is less likely than another collection, it is almost impossible to dissuade them by appeals to reason. So if Cedric, the pub bore, offers this opinion, try another line of attack: if reason does not work, losing money might. Borrow the numbers 1 to 5 from the set of pool balls, and place them in a closed bag. Let him draw three balls at random—there are just ten outcomes (Appendix I). To be consistent, Cedric will agree that {1 2 3} is less likely than average, so he will give you better odds on that outcome. Take up his bet. Either he will see the light, or you will become richer.

Accepting that all Lottery combinations are equally likely, we must count how many ways there are of choosing six numbers from the 49 numbers available. In the notation of Appendix I, this is just $^{49}C_6$, which evaluates as 13 983 816.

With one ticket, your chance of matching all six numbers drawn is one in 13 983 816 or, between friends, one in 14 million. How best to make sense of such a figure? Chances such as one in two, or one in ten, are well understood, as we can relate them to everyday experiences in tossing coins, or whether the car might fail its MoT test. When chances are described as one in a million, or one in a billion, the only message that is received is that the chance is remote; little distinction is seen between them, even though one is a thousand times as big as the other.

Imagine spending £1000 *each week* buying Lottery tickets. You would then, on average, match all six numbers about once every 270 years. Alternatively, some scholars have argued that Stonehenge was a primitive computer. Take a flight of fancy, and suppose that they were correct, but its real purpose was to generate random numbers for the Druid's Lottery. Suppose you and your ancestors had bought 50 tickets every week since the first draw, live and with music, on Salisbury Plain 5000 years ago. With average luck, your family would have shared only *one* jackpot since the beginning! Your ancestors might have spent this £50 every week, over the rest of the Stone Age, the whole of the Bronze Age and the Iron Age; through the Roman occupation of Southern Britain, the Middle Ages, under all those Tudors and Stuarts, and with George after George; from before recorded history, through all the Pharaohs, the Ancient Olympics, through the times of 300 Popes—and only one jackpot to expect. Of course, your family would have won many lesser prizes during the wait, but the chance of matching all six on any one ticket is very, very small.

To find the prospects of winning the lesser prizes, we count how many tickets match three, four or five numbers. Call the six numbers that the machine selects the Good numbers, and the other 43 the Bad numbers. (We will deal with the Bonus number later.) We want to know: if we select

Table 2.1 The numbers of combinations that match 6, 5, ... , 0 of the winning numbers in a 6/49 Lotto.

Number matched	Number of tickets
6	1
5	258
4	13 545
3	246 820
2	1 851 150
1	5 775 588
0	6 096 454

six numbers altogether, how many will be Good, and how many Bad? This is a standard counting problem, solved by the methods of Appendix I.

To illustrate, consider how to win a Match 4 prize. With exactly four of the six Good numbers, we also have just two of the 43 Bad ones. The total number of tickets that win a Match 4 prize is

$$(^6C_4) \times (^{43}C_2)$$

which comes to 13 545. The other numbers in Table 2.1 are found by the same method.

As a check, you should add the numbers in the second column to see that they sum to the total number of different tickets. We paid no attention to *which* six numbers were Good, so this table applies to every set of winning numbers.

Provided all combinations really are equally likely, Table 2.1 contains all the information to find the chance of winning any prize. Just work out what fraction of the 14 million or so combinations qualifies for the prize. Continuing the illustration with the Match 4 prize, that chance is

$$13 545/13 983 816$$

which is just under one in a thousand. A player spending £5 per week can expect to win such a prize about once every four years.

You will have noticed that the biggest number in Table 2.1 is the last one. This means that the most likely result is that you match *none* of the numbers drawn, and matching only one is close behind. You are not unlucky if you match just one number—85% of all tickets sold do no better!

There is a nice direct way to find the *average* number you match on any

Table 2.2 The chances of winning a prize if you buy one ticket in the UK Lottery.

Prize	Number of tickets	Chance, about 1 in
Jackpot	1	14 million
Bonus	6	2.3 million
Match 5	252	55 550
Match 4	13 545	1030
Match 3	246 820	57
(Something)	260 624	54

one ticket. Each of your six numbers matches a winning number with probability 6/49, so, on average, you match $6 \times (6/49) = 36/49$ numbers. (Some people worry when they see an answer like 36/49, objecting that you can't possibly match 36/49 tickets. True; but this is the *average*. It makes perfectly good sense to say that you buy 1.5 CDs a week, on average.)

To win the Bonus prize you need the Bonus number and five of the six Good numbers. This calculation is now easy. The Bonus number replaces exactly one of the six Good numbers, so just six combinations win the Bonus prize. The 258 tickets in Table 2.1 that match exactly five numbers split into six that win the Bonus prize, and 252 that do not. The chances of winning the different prizes are shown in Table 2.2.

This table spells out the odds. The 6/49 format was chosen after considerable research into similar lotteries, and investigation of what would attract gamblers. There are two crucial ingredients: a reasonable chance to win *something*, however little, and the fact that life-changing sums are seen to be won. Someone spending £5 a week can expect about five wins a year, frequent enough to keep their hopes alive. For £1 a week, you can permanently indulge your dreams of what you will do when you win the jackpot. (It is always 'when' you win, not 'if'.) Because the chance of winning the jackpot is so small, its size can be enormous. The regular creation of millionaires inevitably leads to press publicity, and thus more sales. So how much might we win?

Only 45% of money staked is returned as prizes. This means that, however the prizes are structured, and however many tickets are sold, the average return to gamblers is 45p for every £1 they spend. Based purely on this average return, the Lottery is a dreadful bet. But the average arises from a 98% chance of winning nothing, combined with small chances of quite large amounts: the prospect of a giant win can far outweigh the poor average return. Punters are attracted by a chance, however small, to win a substantial fortune. The more tickets that are sold, the more the maximum prize can be. To make the arithmetic for the prizes easy, suppose 100 million tickets have been sold, generating a prize fund of £45 million.

When the Lottery draw is complete, the number of tickets that have matched exactly three numbers is counted. Each of them wins the fixed amount of £10. On average, this takes about 39% of the prize fund, but it might be rather more, or considerably less. Not until all the tickets are checked can we know for certain. For simplicity, assume that the average, £18 million, goes to these fixed prizes. The sum left, £27 million, is called the pools fund.

This pools fund is split in fixed ratios among the other four prizes, with the lion's share going to the jackpot. In our scenario, this puts about £14 million in the jackpot pool. How much each winner gets depends on how many other winners there are, as each of the four pools funds is shared equally amongst the corresponding winning tickets. If only one jackpot-winning ticket has been sold, its holder takes all £14 million; but if ten such tickets have been sold, each winner gets a less spectacular £1.4 million. With 100 winning tickets, you are down to £140 000 each.

You don't think that could happen? Ask any of the people who bought the winning selection for Draw 9 in January 1995, when the jackpot pool was about £16 million! Only a month earlier, a Blackburn factory worker had been the sole jackpot winner in Draw 4, scooping the entire £17 million. But in Draw 9, 133 people had bought the winning combination, so each won a mere £122 510. A worthwhile sum maybe, but how disappointing! If you want to win a very large amount, avoid selections that are so popular that many other people repeat them. Suggestions for how to do this come later.

Apart from the fixed £10 for matching three numbers, you cannot be sure how much you will win. But we can calculate the average value for each prize category, which is the same in a normal draw no matter how many tickets are sold. We can also see how far from this average the prizes have strayed.

Sometimes the jackpot pool has been boosted by a substantial sum. This usually occurs because no ticket had matched all six winning numbers in the previous draw, so that jackpot pool has been rolled over. At other times, Camelot, the Lottery franchisees, have added an extra sum from their reserves to give a so-called superdraw. In either event, the lesser prizes are not affected. The Lottery regulator noticed that it is theoretically possible for there to be so many Match 3 winners that the fixed £10 prize

Table 2.3 Sizes of Lottery prizes (figures rounded) over 282 draws.

Prize	Average amount (£)	Smallest (£)	Largest (£)
Jackpot	2 million	122 500	22.6 million
Bonus	102 000	7900	796 000
Match 5	1520	181	7800
Match 4	62	16	164

exhausts the entire prize fund. He ruled that in this case, all winning tickets, even the jackpot, should get an equal share—£8 perhaps. (Cairo will host the Winter Olympics before this happens.)

Table 2.3 shows these average prizes, and the extremes that have arisen in the first 282 draws to September 1998. Unsurprisingly, the biggest jackpot prize was in a rollover draw. The largest prize won in a normal draw was £11.7 million. But the *smallest* jackpot prize was also in a rollover week—Draw 9 described earlier. Without the extra rollover money, that jackpot would have been reduced to £48 500. The biggest potential win to date was when a jackpot pool of £42 million was shared between three winning tickets. But if the winning combination in *that* draw had been {1 2 3 4 5 6}, Camelot reported that there would have been 30 000 winners, each banking only £1400.

It is very striking how much variation about the average there has been. This variability within each prize category is the second significant consideration, after the knowledge that the average return is only 45% of all stakes. Any prize is better than nothing, of course, but you could never have bought even a new Ford Escort with only a Match 5 prize. History says that if you want to win at least £10 000, you can only do so through the jackpot or bonus prizes, and the chance that one ticket captures one of these prizes is about one in two million.

The prizes have been so variable because some combinations are far more popular than others. In any draw, Camelot know exactly how many tickets were bought for every combination. They guard this information like a poker player with a Royal Flush. So far as I know, they have made only two relevant statements:

(1) {1 2 3 4 5 6} is extremely popular (10000+ per week).

(2) The single most popular choice is (or was) {7 14 21 28 35 42}.

Jonathan Simon has reported on gambler choice in Draw 101, October 1996, when each combination was bought five times, on average. The jackpot pool forecast, before the draw, was for about £10 million. There were 2000 combinations that were so popular that had they been the winning set, more than 200 tickets would have shared the jackpot, with each winner receiving less than £50 000. These 2000 combinations represent only one in 7000 of all possible selections, so one of them will be drawn about once in 70 years, if the format remains at two draws per week. Another 200 000 combinations were chosen by more than 20 people.

At the other extreme, in this Draw 101, one-third of all the combinations were bought by at most two people. Indeed, nearly 700 000 combinations, some 5% of the total, and including that week's winning combination, were not chosen by anybody! About 12% attracted just one purchaser, who stood to scoop the entire jackpot, and another 16% of combinations were bought just twice. Plainly, there was a good prospect, before the draw was made, of the main prize being £5 million, or more.

Many people made an initial choice of 'their' numbers that they have clung on to since the Lottery started, and dare not give up now. Players may buy a season ticket that automatically enters the same set of numbers for future draws. This ensures a certain stability to the selections from draw to draw, although the use of Lucky Dip to generate combinations at random helps change the distribution of frequencies each time.

Gamblers would dearly like to know what are the popular combinations, in order to avoid them. Of course, if anyone publicized a single known *unpopular* combination, it would then become extremely popular! Is it possible to work out for ourselves which are the unpopular ones?

A winning strategy?

I can offer some pointers. They are mainly towards what to avoid, and carry the inevitable 'health warning' that they are based on what people have done in the past. Future behaviour might be very different. If you buy Lottery tickets, have fun, but expect to lose money.

Some other countries make data more freely available. The numbers of gamblers choosing each individual number are published weekly for Canadian 6/49 lotteries. There are some changes from week to week, but the most, and the least, popular numbers remain remarkably consistent. In descending order of popularity, with minor weekly variations, the list is

7 11 3 9 5 27 31 8 17 ... 30 46 38 45 20 41 48 39 40

with the remaining numbers more closely bunched in the middle.

It is tempting to think that the combination {40 39 48 41 20 45} is very unpopular, but that would be quite erroneous. *Because* Canadian lottery players have noted that each of these individual numbers is unpopular, many of them buy that particular set! (This is a beautiful illustration of non-independence. The set of the six *least* popular numbers is one of the *most* popular choices!) But the list does show that low odd numbers are relatively popular, and high numbers less so.

Hans Riedwyl has published extensive and fascinating data relating to one draw in the Swiss 6/45 lottery. This lottery has (Appendix I) $^{45}C_6 = 8\,145\,060$ different combinations, and 16 862 596 tickets were sold that draw, an average of about two for each possible choice. He listed over 5000 combinations that were chosen more than 50 times! When first I saw this, I found it completely astonishing, and I craved access to Camelot's data to see if UK Lottery players behaved similarly. The information above about Draw 101 suggests they probably do. Riedwyl's data are worth exploring further.

The Swiss ticket, shown in Figure 2.1, lays the 45 numbers out as seven rows with six numbers, and a final half-row. Two combinations were each selected by over 24 000 people! They were those running in diagonal straight lines from NE to SW, and NW to SE at the top of the ticket. Every full row of the ticket was chosen over 2500 times, as were many blocks of six consecutive in a column, and other diagonals. Pretty geometric patterns also attracted many buyers.

Why did over 12 000 people buy the apparently random combination {8 14 17 25 31 36}? It was the winning combination from the previous week! Moreover, nearly 4000 people bought the two tickets which differed from these numbers uniformly by either −1 or +1. And every winning combination for the previous year had hundreds of buyers. Even the winning combinations from several years ago, and from recent

Figure 2.1. The layout of the 6/45 Swiss lottery ticket.

lotteries in Austria, France and Germany were highly popular. In short, if there was a *systematic* way to select the six numbers, many Swiss lottery players used it.

This is a widespread phenomenon. In the Irish lottery, two gamblers who shared a jackpot both chose their numbers from the dates of birth, ordination and death of the same priest. When Pope John XXIII died, many Italian lottery players used the time and date of his death, and his age. If you set out deliberately to construct your choices by some logical process, you run the risk of falling into the same thought pattern as some of the other 30 million players. To avoid this you might choose your numbers completely at random—if you have no idea what combination will result, neither has anyone else! Of course, if your 'random' choice did happen to be {1 2 3 4 5 6}, or any other combination that you have good reason to suspect is popular, you would do well to reject it and select again.

Humans find it hard to make genuine random selections without invoking some device such as coins, dice or playing cards. Left to themselves to dream up random numbers, they select some numbers, and some types of combinations, far too often, others very seldom. We can only guess the extent Canadian gamblers thought they were choosing at random when their selections led to the number order shown above. The relative rarity of the 'round' numbers 20, 30, and 40 does suggest a subconscious desire to deliberately construct numbers that 'look' random. Try the simple experiment of asking friends to write down two numbers, at random, from the list 1 to 10. Unless they have read this book, very few of them will select 5, and most of them will choose a pair that differ from each other by a middling amount, three, four or five.

New Zealand has had a weekly 6/40 Lotto since 1987, with recent sales of about 14 million in a normal draw. This leads to an average of about 3.7 purchases of each of the four million or so combinations possible. One big difference in gambler behaviour is in the use of Lucky Dip. In the UK, after a slow beginning, the use of Lucky Dip increased, and accounts for some 15–20% of all sales, and more at rollover times. In New Zealand, a massive 62% of tickets are bought using this 'official' method of generating random selections, with only 38% bought on individual whim—birthdays, lucky numbers, deliberate choice of previous winning sets, and the like. One consequence of the popularity of Lucky Dip is that the numbers of winners in each prize category are much less variable from week to week, and so the sizes of the prizes are more predictable. At first sight, that sentence may seem odd: the *more* gamblers are choosing at random, the *less* variable the outcomes. Greater use of randomness leads to greater predictability!

There is a statistical explanation for this. We saw above that, left to their own devices, gamblers have a tendency to avoid some numbers and combinations, but select others much more often. Human choices are *very* variable, about any given average: quite a few combinations are almost never chosen, while others are seen to be extremely popular. Computers, that lack wit, intelligence, imagination and subtlety merely carry out instructions at great speed and with total accuracy. These instructions—a computer programme—give selections as close to genuine randomness as makes no difference. And *this* randomness is much less variable about the average than humans collectively manage.

The accumulated data give clues to what choices have been popular in the UK Lottery. To estimate the popularity of any single number, you can count up the number of prize winners when that has been one of the winning numbers. Then compare this with the expected number of winners, taking account of the level of sales. The more excess winners compared with the average, the more popular that number. Several investigators have followed this path, or used more sophisticated methods, and their conclusions are reasonably consistent. The consensus of the answers differs a little from the Canadian data, but not dramatically so. You will not go far wrong if you assume that the relative popularity of the individual numbers in the UK Lottery follows the list given above for Canada. But knowing how popular each number is does not necessarily help—we need the popularity of *combinations*.

The most striking event in a single UK draw was the 133 jackpot winners in Draw 9. Based on the number of tickets sold that draw, five winners might have been expected. That actual result is a complete statistical freak, *if* gamblers are selecting at random. That draw alone is convincing evidence they are not. But to appreciate why so many made that selection, you should remember how the UK ticket is laid out, as in Figure 2.2. Mark the winning combination, {7   17   23   32   38   42}, on this layout, and see that these six numbers:

(1) are all on different rows

(2) are only in the middle three columns

(3) have no pair that differ by exactly five, and so are next to each other vertically.

1	2	3	4	5
6	7	8	9	10
11	12	13	14	15
16	17	18	19	20
21	22	23	24	25
26	27	28	29	30
31	32	33	34	35
36	37	38	39	40
41	42	43	44	45
46	47	48	49	

Figure 2.2. The layout of the 6/49 UK Lottery ticket.

It is as though the buyers moved their pencils down the middle of the ticket, dodging from side to side a little, possibly believing they were choosing numbers at random. (I find this explanation more plausible than that offered by a newspaper at the time: that many Lottery players were born in the years 1938 and 1942!)

One draw cannot be conclusive. There are other explanations for why that combination was so popular. So what happened on the other occasions that the winning numbers fell into this same pattern? We must take into account the level of sales. In a weekend draw, some 56 million to 70 million tickets are sold, giving an average of four to five for each of the 14 million possible combinations. In a midweek draw, this average drops to about two. At the time of writing, the pattern we have identified above has been a winning combination on three other occasions. There were 57 and 20 winners in two weekend draws, and 16 winners in a midweek draw, considerably above average every time. To date, there have been no other occasions where there have been more than 20 who shared the jackpot. I may be jumping to the wrong conclusions, but I suggest you should *avoid* buying tickets that fall into this pattern, if you want to win a large jackpot.

Tickets with this pattern tend to have their six numbers fairly evenly spread. One possible measure of how evenly numbers are spread is to ask how far apart are the closest pair. In Draw 9, the closest pair is 38 and 42, which are four apart. Call this value the *minimum separation*, M say. When

Table 2.4 The first 282 draws of the UK Lottery. The "minimum separation" is
M, as described in the text. The totals of the actual numbers of jackpot winners,
and the average expected given the level of sales, are compared, and the impact
on the average size of the jackpot prize is assessed.

Minimum separation	How often	Actual number	Average number	Size of jackpot
1	132	330	514	Higher
2	83	312	339	Higher
3	31	144	113	Lower
4	22	259	84	Lower
5 or more	14	146	52	Lower

there is an adjacent pair of numbers, then $M = 1$. Fairly obviously, M lies
between one and nine, and the larger it is, the more evenly scattered the
selections tend to be (all agreed?).

We can gain insight into gambler choice by comparing the actual
number of jackpot winners with the average number we would expect,
given the level of sales, as described above. For each draw, we know the
actual number of jackpot winners, and this average number. So to see what
has happened overall for a given value of M, we add together the results for
all the draws when this occurred.

In 132 of the first 282 draws, the winning combination had an adjacent
pair of numbers, i.e. $M = 1$. Just looking at those 132 draws, a total of 330
tickets shared the jackpots. On average, given the sales, as many as 514
would be expected. *There have been far too few winners, overall, when the
winning set contained an adjacent pair.* Consequently, the prize for winning
the jackpot has (usually) been higher.

Similar calculations for the other values of M give Table 2.4. Focus on
the third and fourth columns to see whether the actual number of jackpot
winners is more or less than the average expected, given the level of sales.
The table shows that when the numbers are more evenly spread, e.g. when
the minimum separation is at least three, there have been more winners
than average, so their prizes have usually been lower. But do treat this table
with some caution: the last two rows are based on very few draws, 22 and
14 respectively, so the picture could change as more data arise. And
remember that, although the combination {1 2 3 4 5 6} has a minimum
separation of just one, it would undoubtedly yield a disastrously low prize!
This criterion is too simple on its own to capture more than a hint of a
good choice of ticket.

We can use the same logic for other criteria, apart from this minimum separation. Do gamblers choose high numbers often enough? (What is a high number? How long is a piece of string?) If we take the high numbers as 41 to 49, which occupy the bottom two rows of the UK ticket, the message is rather curious. There are *fewer* winners than expected when the winning ticket has none of these high numbers, and *more* than expected when there is exactly one. But when there have been three or more high numbers, there have been *very* few winners. It seems that gamblers like to include some number from the bottom two rows, but chary of including three or more. So far, to Draw 282, the winning combination has never had five or six numbers in the bottom two rows, so I can only speculate on the likely outcome. But knowing that the combination of the first six numbers is very popular, I would risk a modest wager that the combination of the last six numbers has many adherents, so {44 45 46 47 48 49} could well yield a disappointing prize.

But future behaviour may be different.

More than one ticket?

The more tickets you buy, the more chance you have of winning the jackpot. To an extent, buying many tickets for the same draw is like backing several horses in the same race—you are betting against yourself. However, unless you are contemplating buying half a million or more tickets, it makes no real difference whether you buy them all for one draw, or spread your purchases over many draws.

How can we see this? For any one ticket, write p as your (tiny) chance of winning the jackpot. Table 2.1 shows that $p = 1/13\,983\,816$. If you buy N different tickets in the same draw, your chance of winning the jackpot is exactly Np.

With any one ticket, your chance of *not* winning the jackpot is $1 - p$. If you buy one ticket in each of N draws, the chance you never win is $(1 - p)^N$. You either never win, or win at least once, so your chance of winning at least once is

$$1 - (1 - p)^N$$

This answer is different from the previous one, Np, but not by very much. So long as N is less than half a million, the two answers are within 2% of each other. Indeed, for N less than 1000, they are within one part in 25 000.

If you buy a total of N tickets, unless N is ridiculously large, you can distribute your purchases among as many draws as you like without noticeably affecting your chance of a jackpot win. Your average return is also the same. Avoid carelessly buying the same combination twice for the same draw.

Perhaps this advice will cause raised eyebrows with the pub syndicate in Kent who, in September 1997, won nearly £11 million when they won two of five shares in a double rollover jackpot. They were very, very fortunate! Their intention was to spend £28 to buy all combinations of six numbers from a set of eight. However, the man entrusted with filling in the tickets made a slip: instead of 28 different combinations, he entered only 27 of them, but one of them was duplicated. By great good fortune, that duplicated ticket was the jackpot-winning combination! The syndicate won two jackpot shares, 12 Match 5 prizes, and 14 Match 4 prizes through this overlap. It could so easily have been a disaster—instead of omitting what turned out to be a Match 4 winning ticket and duplicating the jackpot, they might have duplicated a Match 4 ticket and not bought the jackpot ticket at all!

Each jackpot share was worth £5.4 million, so two shares brought twice that. The dozen Match 5 prizes added just another £24 000. Had they bought the 28 different tickets intended, the jackpot would have split just four ways, so their one-quarter share would have risen to £6.8 million. But if no other people had bought the jackpot-winning combination, the syndicate would have got exactly the same reward whether they had bought one jackpot ticket or two! If one other winning ticket had been bought, they would have got 2/3 of the pool. You cannot know in advance which combination will win, and how many others will select it, so buying different tickets is a better *strategy*. Two chances of winning half a jackpot are superior to one chance of 2/3 of a jackpot.

The popular press enjoyed the human story of the Kent syndicate's good fortune, but quoted the curious odds of '196 million to one' against the event. It is interesting to speculate on how this implausible figure was calculated. I would argue that a syndicate entering 28 different combinations has 28 chances in 14 million of winning a jackpot share. If it enters 27 combinations, one of which is duplicated, its chance of some jackpot share is 27 in 14 million. And the chance the duplicated ticket wins is simply 1 in 14 million. End of story. Nowhere do the quoted odds arise. My belief is that the figure came through two errors: first, someone argued that as the chance of one win is 1 in 14 million, so the chance of two wins 'must' be that figure *squared*; and then compounded this error by squaring 14 million and not appreciating that the answer would then be 196 million *million*!

In general, if you do buy several tickets for the same draw, you should consider the extent to which their numbers overlap. If your tickets have several numbers in common, and one of them wins a prize, some of the others may well also be winners. You could buy a Lottery wheel, a device to help you select combinations restricted to your favourite 15 (say) numbers. Such a wheel has properties rather like Plans used by punters on the Football Pools (see the next chapter), but there are good practical reasons to use a Pools Plan. Express a healthy scepticism when anyone enthuses about the 'benefits' of a Lottery wheel. Wheels are hyped on the premise that they help you to win more prizes. Up to a point, Lord Copper. It is indeed true that if you use a wheel and win at least one prize, you are more likely than average to win another—but this is balanced by the lower than average chance of winning any prize at all! Overall, on average, the number of prizes is exactly the same, however you make the selections, for a given number of tickets. Your chance of winning the jackpot is also exactly the same, provided only that you take care to make all your combinations different. If you did win a large jackpot, getting a few more minor prizes (such as won by the Kent syndicate) is the least of your concerns. Averages are wonderfully direct ways of reaching the heart of an argument.

Are the numbers drawn really random?

All the above has been based on the supposition that all 14 million combinations have the same chance of being chosen by Camelot's machines. I have argued that it is rational to *believe* this to hold, before we see the outcomes. But how do we know that Camelot are not cheating? Maybe some balls are secretly heavier than others, or have magnets inside them, or televisual trickery imposes pre-selected images on the balls that are chosen? Is there some giant conspiracy to hoodwink the public? One counter is to point out that it is far less trouble to run a fair Lottery than to introduce bias, and have to guard against being found out. However, if some combinations really did arise more frequently than others, our analysis would be quite different.

A finicky observer might point out that no man-made machine can be so flawless that all the numbers have *exactly* the same chance of arising. This is technically correct, but of no practical importance whatsoever. All that is necessary is that there is no rational reason to believe that some particular combinations are more likely to arise than others. Judging whether the Lottery is behaving as expected is similar to deciding whether the dice in your Monopoly or Backgammon set are truly random, or whether a coin used in the toss at cricket is fair. The logic behind how to come to a decision is identical; the details differ. See Appendix IV for a description.

With 14 million possible combinations, it is impracticable to get enough data from the actual draws to see whether all these combinations have equal chances. (But Lucky Dip is claimed to be a computer-generated random selection of six numbers. Over 10 million Lucky Dip tickets are sold each week, so it is quite feasible to test Lucky Dip on all combinations.) On the limited data from the draws, we must be less ambitious. The two main considerations might well be:

(1) Are all individual numbers equally likely?

(2) Does one draw's result give any clue to the next draw?

As more data arise, we can look at other factors, such as whether all pairs of numbers appear equally often within a draw. Even without using the 'proper' statistical procedure of the Appendix, we can make an initial judgement fairly easily.

Suppose we have data from D draws to judge whether all numbers have the same chance of arising. Each draw gives six (main) numbers, so there are $6D$ numbers drawn. The average frequency for any one number is then $6D/49$—call this A. Some numbers will have appeared more than this, others less often. For example, by September 1998 after 282 draws, each number had come up $A = 34.5$ times on average. Even if some numbers had never been drawn, and others had come up over 50 times, the *average* would still be 34.5. But people's notions of how close to average it is reasonable to expect data to fall are not well developed. I will offer two tests as a rough guide.

Work out the square root of this average, and round it down a little. When $A = 34.5$, its square root rounds down to 5.5. This square root will serve as proxy for what an engineer might call the *tolerance*. We have a list of 49 data values, and this tolerance is the key to whether they are reasonably close to the average, or suspiciously far from it.

The first test is to check that about two-thirds of them are within *one* tolerance of the average. Here, this corresponds to the range from about 29 to 40. We expect about 32 of the 49 frequencies to fall in this range, and it would be a danger signal if no more than 25 of them did so. In fact, exactly 32 frequencies fell in this range. The other test is based on how many frequencies are more than *two* tolerances away from average; up to 5% is fine, but rather more than that is a bad sign. So here we count how many frequencies are below 24, or above 45; we hope there are no more than three. In these first 282 draws, no frequencies were below 24, and only one exceeded 45. This second test is also passed very comfortably. The scatter about the average looks consistent with all the numbers being equally likely.

For a pukka statistical test, we prefer a single overall measure of how close to average these 49 frequencies are. Appendix IV describes the relevant measure, W, the *goodness-of-fit statistic*. The average value of W for a fair Lottery is 48; values less than 48 indicate the frequencies are even closer to their average than we might expect, whereas if W is much bigger than 48, some frequencies will be suspiciously far from average. Tables tell us how big a value of W is statistically acceptable. A truly fair Lottery would give a value above 65 only one time in 20, a criterion often used in statistics to check this goodness-of-fit. After 282 draws, the figure was 52.4, completely in accord with an unbiased Lottery.

The numbers in the statistical tables of Appendix IV apply to any sufficiently large number of draws, say 40 or more. So if you wish to assess the fairness of the Lottery over the period from draws 200 to 300, you can use the description in the Appendix to calculate the goodness-of-fit statistic, and use these tables. But remember what you are doing: you are asking what is the chance that a fair Lottery would give such an extreme result. One time in 20, a perfectly fair Lottery *will* lead to a value of W that exceeds 65, and you may wrongly accuse it of being flawed! If you make separate tests on many blocks of draws, sheer random chance will throw up large values of W from time to time. This point is taken up in a different context later.

You will notice I have not gone so far as to assert that the Lottery is definitely fair. Statistics can never *prove* the Lottery is fair, or unfair, in the way that mathematics can prove that Pythagoras' theorem applies to all right-angled triangles. When W is acceptably moderate, the statistical response is not 'Hurrah, the Lottery is fair', it is the more accurate 'There is no evidence it is unfair'. It is rather like a jury in a court of law: they are not asked to decide whether the accused is innocent, they are asked whether the evidence demonstrates guilt beyond reasonable doubt. If it does not, the verdict is 'Not guilty', even if the jury were 75% sure the accused committed the crime. Seventy-five per cent is not enough. In testing Lottery frequencies, there will be no final verdict at any stage, except that if W became so large that public confidence in the Lottery's fairness were dented, some new method of generating the numbers would surely be found.

This goodness-of-fit test is a jack of all trades, in the sense that if there are any differences at all in the inherent probabilities of selecting the numbers, the test will pick them up, eventually. But because it is a portmanteau test, it may take some time to detect any particular differences. If there were discrepancies of about 2% in the frequencies of the individual numbers—quite enough to cause concern—this test has only a slim chance of detecting such differences over as many as 300 draws. Look at the example in Appendix IV with a hypothetical coin that falls Heads 47% of the time to see how hard it can be to detect appreciable bias.

To have equal frequencies is not enough. If the Lottery cycled regularly through all the 49 numbers in turn, the frequencies would be as nearly equal as possible, but such a cycle would be completely unacceptable. The unpredictability of future draws is even more necessary than equal frequencies for the numbers.

Some Lottery pundits offer charts and tables, with the intention that these past data can be used to infer future behaviour. One school holds that numbers that have just appeared are 'hot' and are therefore more likely to be drawn again. The rival 'LoA' school asserts that 'the law of averages' implies that numbers that have *not* appeared recently must be drawn sooner rather than later, so as to help equalize long-run frequency. (Neither school explains *how* numbers are able to remember their history, nor how they could be capable of making extra efforts to appear when the charts suggest they are due.) Common sense demands that past draws are irrelevant to the future, but common sense can be wrong. How can statistics indicate whether either school has a point?

Suppose that, for each upcoming draw, you buy one ticket consisting of the *last* winning combination. If the 'hot' school have a valid point, you should match *more* numbers than average; using the LoA argument, you ought to match *fewer* than average. If successive draws really are independent, you should match just as many as if you bought a random ticket.

Once again, simple averages reach the heart of the matter. Buying one ticket, the average number of numbers you match is 36/49 (we worked this out earlier). So buying one ticket from each of the most recent 281 draws, you should match 281 × 36/49 = 206.4, on average. Anyone who had operated this 'play-the-winner' strategy of buying the winning combination from the previous draw would actually have matched 190 numbers altogether in this period. You do not need a degree in statistics to appreciate that 190 is sufficiently close to the average of 206.4 to be judged consistent with what should happen if the draws are independent. There is no evidence to support either alternative 'school'.

This analysis uses data from only two consecutive draws. Perhaps some lack of independence only becomes apparent when data from longer sequences are used? To investigate this possibility, let's look at what happens when some number seems to be occurring unusually often. For definiteness, call a number 'hot' if it has appeared at least three times in the previous seven draws. Thus some draws will have no hot numbers, others may have several. Most of the time, there will be one, two or three hot numbers in a given draw.

We need a history of seven draws to begin the investigation, so look at all the subsequent draws from Draw 8 to Draw 282. They contained respectively 2, 1, 3, 2, ... ,3, 2 hot numbers, a total of 557. We dull sceptics, who stubbornly invest the numbers with no magical powers to force their way forward, or hide at the back, expect these 557 numbers to do no better and no worse than average. 'Average' here corresponds to any number having chance 6/49 of arising in any given draw. Thus average behaviour is 557 × 6/49 = 68.2 appearances. The LoA school expects rather fewer, the other school expects rather more. In fact, these numbers arose on cue just 67 times—as close to average as one could dare to hope. (Any actual frequency of between about 52 and 84 would have passed standard statistical tests.)

The choice of three appearances in seven draws was quite arbitrary. I really had no idea what the outcome would be when I selected it. I might equally well have selected something quite different as the criterion—perhaps three from six, or four from ten, or two from five. And this leads to an important, but subtle point, hinted at earlier.

At any time, we have a given quantity of data. There are many possible criteria for a 'hot' number, and it would be remarkable if all of them gave such a good match between the actual result, and the average expected from independence. For a good analogy, ask how suspicious you should be if a coin falls Heads ten times in succession. If those were the only ten tosses made, you would be right to be very suspicious. But if the coin had been tossed thousands of times, and somewhere in the sequence there were ten successive Heads, no valid suspicion would arise. Indeed, if the sequence is long enough, it would be fishy if there were *no* set of ten consecutive Heads! So it is with the alternative definitions of 'hot' Lottery numbers: if you try enough alternatives—four appearances in the last 12 draws, two from three, keep on looking and compare actual with average each time—by sheer random chance you will, occasionally, find an outcome 'unusually' far from average.

Suppose you do trawl through the data and find one of these apparently anomalous outcomes. How can you tell whether the anomaly is real? The fair test is to see whether the same phenomenon arises in future draws. It is not scientifically valid to use the same set of draws both to *discover* unusual behaviour, and also to 'prove' the draws are not behaving randomly. The Lottery is so rich in data that it is inevitable that some outcomes that seem inherently improbable have arisen. Here are three.

About a year after the Lottery had started, and some numbers had come up as main numbers ten or more times, the then laggard, 39, had appeared only once. Some saw this as virtual proof that the Lottery was flawed! Statistics is more prosaic. If you name a number, then the chance that it does not arise in (say) 50 draws is very small—about 0.0015; but the chance that *some* number makes no appearances in 50 draws is about 7%—not large, I agree, but no longer absurdly small. Moreover, the number 39 was only identified as a possible rogue when it had failed to appear during about 40 draws. The true test of it having a lesser chance of appearing is to start counting its absences *after* suspicion had been thrown on it. (39 did appear again, of course, and after a total absence of 52 draws. Random chance eventually moved it off the bottom of the frequency table, and random chance took it back again. After 282 draws, the numbers 37 and 39 jointly had the lowest frequency. The 'record' absence of 52 draws was overtaken by the number 15, which was drawn in Draw 100, and then not again until Draw 166. That record too will fall sometime.)

Newspapers published correspondence from readers who had noticed that during a particular sequence of 37 draws, some *triple* (i.e. a set of three consecutive numbers, such as {16, 17, 18}) had appeared in the winning combination eight times. One reader asserted that this was stretching the laws of probability to breaking point! Again, care is needed. The chance that the next 37 draws contain eight or more triples is indeed very small— under 1 in 3000. But the 37 draws that gave rise to this correspondence were not selected arbitrarily. The first and the last in the sequence were included precisely because they marked the boundaries between apparently few draws with triples, and this surfeit. Over all draws to date, there is no real evidence that triples arise more often than random chance would give.

A final example of an inappropriate *post hoc* analysis arose in the late summer of 1996. The number 44 was (astonishingly?) drawn eight times in the ten draws from Draw 86 to Draw 95. To be strictly honest, it was the Bonus number in two of these draws, but if it helps make a good story, details like that tend to be overlooked. Again, the pertinent question is: given all the draws we have had, should we really be surprised that one of the 49 numbers has turned up that frequently, at least once? (Especially as we can choose to include or to exclude the Bonus number as the mood takes us, if it helps make the event 'look' more non-random.)

All three examples would have justly carried the tag of 'worthy of attention' if they had been flagged in advance. But they were not. There are plenty of 'unusual' things that might have happened, but have not. So far, to draw 282, no two consecutive draws have had four or more numbers in common—Table 2.2 shows this happens about one draw in 1000. It took until draw 267 to have all six main numbers odd, although that will occur about once in 79 draws in the long run. We could dream up thousands of things. If we look back at the data, some of them will have occurred rather more often than we might expect, others too infrequently, most of them about average.

Do not get too excited if you unearth one of these 'anomalous' events; start your observations from now, to see whether the anomalous behaviour continues. Perhaps it will, and you think you have detected a feature in the Lottery that will improve your chances of winning. Go ahead, use it. You will do no worse than average, in the long run. As a cricket captain, I used a method passed on to me by the seventh son of a seventh son: I called Heads on grass, and Tails on all other surfaces. I won 50% of the time—and you can't do better than that!

Summary

The UK National Lottery 6/49 game has the same format as similar games in other countries. The main object of all these games is to raise money from gamblers, so the amount returned as prizes is much less than is spent.

In the UK, prize money is 45% of stakes. But the £1 for a ticket buys several chances of substantial prizes, and the Lottery has created hundreds of new millionaires. In addition, holding a ticket, anticipating the chance of being lucky, has value in itself.

The chances of winning the different types of prizes are shown in Table 2.2, and Table 2.3 gives information on the size of prizes. Ninety-eight per cent of all tickets bought win nothing, and 95% of all prizes are the fixed prize of £10. Only the jackpot (one chance in 14 million) and bonus (one in 2.3 million) prizes are likely to be large enough to change a lifestyle. Some gambler choices have been far more popular than others, and if such combinations are drawn, even these maximum prizes will be disappointing. Information from Lotteries in other countries as well as the UK can help you avoid these popular selections, and so lead to larger prizes (if you win). The popularity of choices may not remain fixed over time. Choosing numbers genuinely at random protects a gambler from repeating the thought patterns, and perhaps the choices, of other punters. Buying tickets only when the jackpot is boosted by a rollover or a superdraw increases the average return above 45%; however, the lesser prizes are not increased, so unless all six numbers are matched, this extra 'free' money is irrelevant. For a given (reasonable) amount spent, it makes no noticeable difference whether tickets are bought for one draw, or spread between several.

There is no evidence pointing to any single numbers, or combinations of numbers, that are more or less likely to be drawn by the Lottery machines than others. On the other hand, if some numbers did have a greater chance of selection, that might well take many years to become apparent.

Postscript (November 1999)

Table 2.3 is unchanged for regular draws, but the bonus prize was nearly £2 million in draw 286, when extra free money went into the bonus pool. The broad message in Table 2.4 is the same, but draw 366 illustrates its limitations: 46 tickets shared the jackpot, even though two winning numbers were adjacent. That same draw, since only 13 shared the bonus pool, the bonus prize exceeded the jackpot prize. Five draws later, no bonus prize was won, so the bonus pool was added to the jackpot pool, giving the two winners an extra £650 000 each. "Unto every one that hath shall be given"—St Matthew 25(29).

CHAPTER 3

Football Pools, Premium Bonds

Football Pools

In 1923, three telegraphists from Manchester rented offices to launch the first Football Pools competition. Fearing their current employers would disapprove, they disguised their identity through an obscure family link, and used 'Littlewoods' as a cover name. They distributed 4000 coupons to fans queuing to watch a Manchester United game, and just 35 people entered that first week. A few weeks later, only one coupon from 10 000 handed out to soccer fans in Hull was returned. At the end of the first season, the venture had made a loss, and two of the original partners dropped out, but John Moores carried on. Despite this unpromising beginning, the idea survived, and today ten million people play the Pools regularly. Although rival organizations have run substantial competitions on similar lines, Littlewoods dominates the market and offers the largest prizes.

By 1928, the prize pool had reached £10 000 a week. Increased sales and inflation lifted the size of the top prizes, and a six-figure sum was paid in 1950. At this time, when soccer had a maximum wage of £10 a week, and the Beatles were still in short pants, the standard way for ordinary people to aspire to riches was through 'the Pools'. If you had selected the only eight drawn matches one Saturday, you could expect a prize of many life-times' earnings for a stake of one penny. The odds, long even by Lottery standards, were largely ignored.

To maintain interest over the British summer, matches from Australia have been used since 1951. When Arctic weather wiped out large numbers of matches in 1963, a 'Pools panel' was set up to invent results for postponed matches, so that the weekly gamble could continue. The first payout of half a million pounds was in 1972 and, seven years later, an unemployed hairdresser spent 45p on her first ever Pools entry, winning over three-quarters of a million pounds. A group of nurses from Devizes, who asked their patients to select a set of numbers, shared a million pounds in 1984, and a single winner in 1991 received over two million pounds, for a stake of 54p. The Lottery has seriously affected the Pools business, but very large prizes are still won for minuscule stakes.

Today the Pools firms have abandoned the Penny Points, the Easy Six, and complex scoring systems with half points here and there. The main challenge on Littlewoods' coupons is the 'Treble Chance', where a punter hopes to select eight soccer matches that will lead to a score draw from a list of 49 games. The original Treble Chance, so called because it paid out three dividends, began in 1946. Now each score draw is worth three points, a no-score draw gains two points, and a home win or away win is worth one point. For added interest, each entry is checked twice, once on the final results when the biggest prizes can be won, but also on the half time scores.

The prize money paid out is what is left after government tax, a contribution to the Football Trust, and the firms' expenses have been deducted. The proportion varies from week to week, but on average less than 25% is returned to gamblers. In a typical week, about four million coupons are sent in, representing £8–9 million staked, and £2 million paid out as prizes. About an eighth of the prize fund is provisionally set aside for the competition based on the half-time scores. If no entry obtains the maximum points achievable on that week's half-time scores, this provisional sum enhances the dividends on the full-time results. The possibility of a gigantic payout arises since three-quarters of the full-time prize pool is allocated to the first dividend. Three lesser dividends are paid from the other 25%. All the prize money is paid out each week, there is no rollover. To illustrate what might happen, the (rounded) figures for three successive weeks in July 1997 are shown in Table 3.1.

These weeks were during the British summer, using matches from

Table 3.1

	12 July	19 July	26 July
First dividend	1 prize of £966 000	263 of £3700	5 of £205 000
Second dividend	6 prizes of £14 500	3165 of £31	324 of £490
Third dividend	207 prizes of £400	30 480 of £2.60	994 of £147
Fourth dividend	975 prizes of £145	155 381 of £1	22 701 of £9
Half time scores	15 prizes of £16 000	8 of £29 000	No client with maximum

Australian soccer: total stakes are normally higher in the winter. On 12 July, barely 1200 prizes altogether were awarded; the next week there were nearly 200 000! The variability in the sizes of the prizes over that short period is a clear warning not to celebrate your first dividend prize too well, until you learn how much it will be. You can make a good guess, taking account of the number of score draws, and the Pools firms quickly offer sound bites to the media—'possible jackpot', 'dividend forecast good/ moderate/low'—as their banks of computers begin to scan the millions of lines entered. But the figures here are worth close scrutiny: the main prize was close to a million pounds one week, but next week's winners would be barely able to fit out a new kitchen, or support one daughter through a year at university. The *second* dividend from one week, and all the half-time prizes, greatly exceeded the top prize on another week. Sometimes there are so many potential winners of the lower dividends that the amount due would be less than £1, in which case that prize money is re-allocated to the higher levels.

The size of the top prize depends heavily on the number of draws. Some gamblers seek to apply skill, by selecting games that seem to be between well-matched teams, or identifying derby matches, but the large majority of Pools players are more passive. It helps very little to pick a score draw that most other punters have also selected. Over 80% of customers either use exactly the same numbers each week, or claim to make their selections entirely at random. The first woman to win over two million pounds kept with identical numbers for 18 years, winning *no prizes at all* before her spectacular scoop. If the score draws are concentrated in matches 1–31, it becomes more likely that the top prize will be shared. There are two reasons for this: the first is the use of 'lucky numbers' associated with birthdays; second, matches with numbers higher than 31 will be from the lower English Leagues, non-League games, and the Scottish Leagues. The large majority of punters are from England, and will probably feel more comfortable making their forecasts from among teams they are more familiar with. But predicting score draws, even for avid followers of soccer form, seems quite haphazard. By and large, gamblers are making selections of eight matches from 49 in a fairly arbitrary way; some deity distributes 3, 2, or 1 points to the list of matches according to His probability law, and this interaction of random events determines the prize winners.

There are (Appendix I) $^{49}C_8$ possible ways of choosing the eight matches. This works out to be just over 450 million. Since each line costs a mere 4/3p, gamblers buy many lines (the average stake is about £2). The

Table 3.2 The numbers of lines, and the corresponding cost when each line costs 4/3p, for a full perm of eight matches.

Number of matches	9	10	11	12	13	14	15
Number of lines	9	45	165	495	1287	3003	6435
Cost (£)	0.12	0.60	2.20	6.60	17.16	40.04	85.80

single most popular entry is to mark ten matches, and to indicate that your entry is all 45 combinations of eight matches from these ten, costing 60p, the minimum total stake permitted. Over 80% of gamblers use similar full perms, by marking more than eight matches and buying all possible combinations of eight from them. Such entries are very easy to check. If the total points from the best eight results are insufficient for a prize, checking is complete; otherwise, the Pools firm can supply you with a table from which you can read off how many winning lines you have. For example, if your ten selections have seven score draws, two no-score draws and a home win, your 45 combinations contain two lines with 23 points, eight with 22 points, fourteen with 21 points and twenty-one with 20 points.

Table 3.2 will help you and your syndicate decide how large a full perm you might use. It gives the numbers of lines, and the total cost at 4/3p per line, for full perms of different sizes. You may stake more than 4/3p for each line, if you wish. If you do this, your winnings are increased proportionately, with the proviso that if you are the sole winner at any prize level, the amount of your prize can never exceed the total available for that dividend.

After full perms, the most popular method of entry is to use an official Plan. Plans are available from the Pools firms, and from some newspapers. Instructions for a Plan might be to mark 16 matches, and write 'LitPlan number 80 = 60 lines at 4/3p = 80p'. The lines of eight matches for this Plan are stored in the firm's computers, making checking automatic. There are Plans to suit all pockets. The most popular newspaper Plan is for 600 lines, costing £8, based around selecting 17 matches. To cover all possible selections of eight matches from 17 would cost over £300. This newspaper Plan constructs its 600 lines in such a way as to guarantee that, if any eight of the 17 matches are score draws, at least seven of them will fall in one of the lines entered. (The mathematics behind such guarantees is quite subtle. It follows exactly the same ideas used in the design of statistical scientific experiments to test new medical treatments, or to judge whether higher crop yields can fairly be attributed to new fertilizers, rather than to random chance.)

These Plans have an affinity with a Lottery wheel (Chapter 2), but there are sound reasons for using them on the Pools. The first relates to the cost of each line. Unless you use a Plan, or some sort of perm, you will have to list every single line separately on your coupon. At a total stake of £2, this would mean marking *and checking* 150 collections of eight matches! The second is that the Plans are efficient ways to utilize any skill you have at picking out scoring draws. The guarantees in these Plans guard against the bad luck of having selected many drawn matches, but not collecting enough of them into the same line. Finally, these Plans (unlike a wheel) are offered free, as the Pools firms can automatically process coupons that use them.

It is not possible to say whether one Plan is 'better' than another. They are convenient ways to make Pools entries at a chosen cost, but the average number of winning lines depends simply on your ability to forecast score draws, and how many lines you buy. On a Plan or a full perm, the overlap between lines means that whenever you have one winning line, you tend to have others, just as with a wheel in the Lottery. Large winners' cheques are usually a composite of one first dividend line, and many lesser prizes.

The term 'jackpot' on Football Pools means a week with a single winner of the first dividend. This happens about once a month, on average, in a quite random fashion. One period of 16 months had no jackpots, but three have arisen in successive weeks several times. Jackpots excite players, and lead to more entries, so the Pools firms would like to set up their games to maximize the chance that one will happen. It is a delicate balance. The firms can use their records to estimate how much money in total the punters will stake in any week. They control the price of each line, which gives them a very good idea of how many lines will be entered. (This explains why the largest firm charges 4/3p per line, while their smaller rivals have used 5/11p, and 1/8p.)

They also decide how many matches will be offered on a coupon. Fifty years ago, about 55 matches were listed, and the scoring system awarded three points for any draw, two for an away win, and one for a home win. Selecting eight matches became firmly established, with many Plans and perms built round that number. Over the years, both the number of matches offered and the scoring system have changed, but two elements seem immutable: punters select eight matches, and the maximum score is 24 points. The number of matches has been as high as 59, and the scoring system has given different numbers of points for no-score draws, 1–1 draws and higher scoring draws. (At one time, the points available were 3, 2 ½ , 2, and 1 ½: this was far more complicated than necessary, as awarding 3, 2, 1, or 0 respectively would have led to exactly the same prize winners!) Now, with 49 games on offer, the score system has returned to 3–2–1, with score draws best, then no-score draws, and home or away wins treated the same.

Their choice of 49 matches means that the total number of different combinations possible, and the total number of lines entered by punters, are of comparable magnitude. On average, each combination is selected about once. (This is a rough guide. By 'about once' here, I mean between about one-third and three!)

Having arranged matters to get each combination selected about once, on average, the Pools firms would now like the number of lines scoring the maximum points available to be small. The number of draws varies between the Leagues, but between one-quarter and one-third of matches are drawn, and more than a quarter of drawn games are goal-less draws. In an average week, there will be about nine or ten score draws, and about four no-score draws. How often will that lead to a jackpot?

When there are at least nine score draws, there are always at least nine different ways to get 24 points (use the second row of Table 3.1), and a jackpot is usually unlikely. With only eight score draws, only one selection gets 24 points, and anyone whose coupon contains a line with those eight matches has a reasonable chance of being the sole winner. But if no-one selects all eight score draws, a jackpot is unlikely. Suppose, for example, there is just one no-score draw that week. Then 23 points can be achieved in eight different ways, by selecting that 0–0 draw with any seven of the eight score draws, so it now becomes likely that there is more than one winner. Each extra no-score draw adds another eight potential winning lines, further decreasing the chance of a jackpot. With only seven score draws, jackpot prospects are better, because now the number of different ways to get the maximum 23 points is just the number of no-score draws, likely to be small.

I will not pursue this analysis to other possible numbers of drawn games. As well as the number of draws, their positions on the coupon will be important in determining how often they are chosen. The Pools firms have kept the familiar features of eight selections and the magic appeal of 24 points, while juggling the prices, the length of the coupon and the scoring system. Striking the right balance to achieve the desired outcome is a diverting exercise. Jackpots do not just grow on trees.

Premium Bonds

When Harold Macmillan introduced Premium Bonds in his 1956 Budget, he said 'This is not gambling, for the subscriber cannot lose.' Church leaders at the time denounced the scheme as 'an encouragement to get something for nothing', and the pithy description offered by Harold Wilson, for the Opposition, was 'a squalid raffle'. Each Bond costs £1, and the government pays a notional rate of interest which creates a flow of money used to pay out prizes on randomly selected Bonds. The first draw, in 1957, offered a modest top prize of £1000, but prizes of £50 000 from 1971, £250 000 from 1980, and £1 million since 1994 have been paid. The odds are long: anyone who has held a single £1 Bond since they were first available is ahead of the game if that Bond has ever won any prize at all.

The rules governing the prize distribution change from time to time, generally after some test of public opinion. The notional interest rate paid tends to reflect the current level of interest rates in the country. In late 1998, that notional interest was 5% per year, and any eligible Bond had one chance in 19 000 of winning some prize in each monthly draw. Bonds can be cashed in for their purchase price at any time, so that an investor exchanges a steady income from interest on money on deposit for a series of chances to win tax-free prizes. Bonds bought in one month are eligible to win prizes from the next month but one. For many years, single £1 Bonds could be bought, but the minimum purchase was raised to 100 Bonds in the interests of 'efficiency'. The extent to which this step has inhibited parents and grandparents from giving Bonds to children at Christmas or birthday time is not known.

In July 1997, the annual interest paid was 4.75%, about 8.4 billion Bonds were eligible for prizes, some £200 for each UK adult, and £33 million was paid that month in prizes. The 439 875 prizes paid were:

- 1 of £1 million
- 5 of £100 000
- 8 of £50 000
- 19 of £25 000
- 47 of £10 000
- 93 of £5000
- 1984 of £1000
- 5952 of £500

- 64 479 of £100

- 367 287 of £50

The full prize structure is described in leaflets in Post Offices, but one major attraction is plainly the monthly prize of £1 million. Each one of the 100 000 people with the maximum holding of 20 000 Bonds has nearly one chance in 400 000, every month, of winning that prize. But three-quarters of all prize money, some 98% of all prizes, goes to the two lowest awards. A single Bond wins at least £1000 only about one draw in every four million. Do not despair at these figures: all you are investing is the net interest you could have earned—probably less than 4p per Bond per *year*.

Bond number 69GK 540100 won the £1 million in July 1997. All Bond numbers have ten positions, though leading zeros are usually suppressed when numbers are quoted. Just as your driving licence number describes your birth date in a transparent code, so a Bond number contains information. In that quoted, the G in the third position indicates that this Bond was one in a block of 2000 sold together. Only the letters Z, B, F, K, L, N, P, S, T or W can appear in the fourth position, and correspond simply to the numbers zero to nine in that order. So that winning Bond is number 693 540 100 in the £2000 denomination range.

The winning numbers are selected by ERNIE (Electronic Random Number Indicator Equipment). The original machine was retired in 1973, and ERNIE Mark III has been used since 1988. Most so-called random numbers used in scientific computation are not random at all: they merely *look* random, and will pass a battery of statistical tests to check whether they are suitable for the purposes to which they are to be put. Such numbers can be reproduced in sequence at will, and the programmer can calculate what value will be generated a thousand steps down the line. Fortunately for the integrity of Premium Bonds, ERNIE generates the winning numbers in a completely different fashion.

ERNIE's operation relies on the really random and unpredictable movement of free electrons in a noise diode. Comparing two such diodes produces a sequence of pulses, according to which of them happens to have the higher noise voltage at any given instant. The number of pulses generated over a short fixed period is counted exactly. This count will be somewhere between 10 000 and 15 000, and just the *last* digit in the count is used. Two such digits determine one of the places on a Bond number.

Given ERNIE's ten separate values for the ten Bond positions, a different computer checks whether that Bond number is eligible to win. Perhaps that number has not yet been sold, or has been sold but repaid, or has already won a prize that month. The first eligible number that ERNIE selects wins the top prize of £1 million, and so on down the list until all the prizes have been allocated. A common fear expressed is 'Has ERNIE left my number out of the draw?'. The answer is that ERNIE is not fed a list of eligible numbers: the machine mindlessly generates Bond numbers that are potentially eligible, and a second machine discards the irrelevant ones. Far from leaving numbers out, ERNIE offers a surfeit.

Despite the faith of engineers that the random movement of electrons will produce a satisfactory outcome, statistical checks on the numbers chosen are also made by the Government Actuary. Details of these checks are published, and the questions addressed include whether:

(1) the individual digits zero to nine appear with equal frequency;

(2) given the current digit, the next digit is also randomly distributed among zero to nine;

(3) taking blocks of four consecutive digits, the 'poker' hands of Four of a Kind, Three of a Kind, Two Pairs, One Pair, or All Digits Different arise at their expected frequencies.

The standard chi-square test (Appendix IV) is used. The final precaution is to check that there is no bias in the order in which the winning numbers are generated. This is vital, as the first numbers drawn win the biggest prizes. One check used for this purpose depends on the code for the third position in a Bond. It is known how many Bonds in each denomination range are held, so for each range, a comparison is made of the actual number of prize-winners at each prize level, against the average expected. Every month, since the scheme began, the Government Actuary has been able to sign a statement that the numbers drawn are suitably free from bias.

Even with all these checks and assurances, scepticism is expressed about the results. One grumble is that more winners of large prizes live in the south. That is true, but then more Bonds are sold in the south. It is alleged that many prize-winning Bonds have two consecutive identical digits in their last six positions: for example, 16 of the first 40 Bonds to win one million pounds have this property. But simple analysis shows this is just what you would expect. There are five pairs of adjacent positions, and random chance will make two given digits different nine times in ten; the chance all five adjacent pairs are different is $(9/10)^5 = 59\%$. This means that the other 41% of the time, the Bond *will* be expected to have an adjacent identical pair—and 16 out of 40 is as close to 41% as you can get. One final dark observation: rather few prize-winning Bonds have a leading zero. This is also true, but many of these Bonds were issued early in the history of the scheme and have been repaid. Bonds cannot be left as gifts in a will, they must be cashed in after their owner dies.

Granted the integrity of the draw, what are your chances of winning, and how much will you win? Plainly, the more Bonds you hold, the higher your chances, but even if you hold the maximum of 20 000 Bonds, fate may treat you cruelly. National Savings sampled 2000 people who held that maximum throughout 1996, when the prize structure was slightly different. The average number of prizes won was 13.6, one person was fortunate enough to win 27 prizes, but one desperately unlucky soul won no prizes whatsoever.

The dominating figure is the average return, currently 5% per year. For anyone taxed at the highest rate, this is as good as a quoted rate of over 8% from a bank or building society, as Bond prizes are tax-free. Actual income from this source is not predictable, but this average return explains why so many well-off married couples both have the maximum permitted £20 000 invested. For the more modest holding of £1000, the *average* annual payout is £50. For each 100 Bond-holders with this holding, 53 will receive no prizes at all during a year, 34 will obtain one prize (usually £50), 11 will get two prizes, and 2 will receive three or more prizes.

Suppose the 100 000 people who held the maximum in July 1997 retained their Bonds for a year. Each can expect to win 12.6 prizes, on average, and the single most likely outcome is 12 prizes. But about four or five of them would win no more than one prize over the whole twelvemonth, and about the same number would find at least 29 cheques dropping on the doormat.

Whatever happens this year has no influence on your fate next year. ERNIE has no idea whether you have already received your ration of winnings. If you seek regular income, find another home for your savings, but each month someone wins a million pounds. It could be you.

The best gamble?

The National Lottery now gives UK residents a third widely available opportunity to win an enormous tax-free prize, albeit at very long odds.

Each of the three gambles has some attraction apart from the monetary prizes. Keen followers of soccer may enter the Pools through their general enthusiasm for the game, and the knowledge that some of their stake money goes to the Football Trust and ground improvements. Twenty-eight per cent of Lottery money goes to charities and other good causes. Bond-holders have to get their inner glow from any empathy with a government desire for a high level of personal savings. In all three cases, your anonymity will be respected if you desire it.

The organizers or promoters of these gambles make it very easy for the punter to participate. To overcome increases in the costs of post, the Pools firms have a network of 70 000 collectors who earn small sums by delivering coupons to punters, picking up entries and stake money, and organizing bulk deliveries. Bonds can be bought at any Post Office, at any time. The Lottery has over 20 000 retail outlets, and punters can even buy a 'season ticket' that automatically enters the same numbers for many draws in advance.

The contests differ in a punter's ability to exercise skill. Bonds, like Lottery scratchcards, have no skill whatsoever. Your money buys the next batch of Bonds in your chosen denomination, just like lifting washing-up liquid from supermarket shelves. Random chance, personified by ERNIE, determines your fate. There is more scope in the Lottery: you can deliberately select a combination that seems to you to be less popular than average, or you can consciously choose Lucky Dip with the same intention. In the Pools, your knowledge of soccer may help you predict which matches are more likely to be score draws. Any ability to pick score draws from the higher numbered matches will tend to increase the average size of any prize you win.

To obtain your Lottery prize, you must make a claim within 180 days of the draw, and present your valid receipt. Camelot will not chase you up; about one million pounds of unclaimed prize money is added to the good causes funds each week. One jackpot prize of over two million pounds was not claimed within the time period, despite Camelot's identification of the locality where the winning ticket had been bought. The Pools firms will send you your prizes, as they have a central record of your entry, but they also ask major winners to make a claim. Holding Bonds, you can be entirely passive, except to notify changes of address. If you have forgotten to do this, and wonder whether some prize is due since 1957, you can obtain a list of all 250 000 unclaimed prizes, adding up to £15 million. Two £25 prizes from that first year still await their owners. There is no time limit on receiving Bond prizes, but National Savings scale down their enquiries after 18 months. For a fee, some private commercial organizations will scan the list of on your behalf. These extra factors will count more in some minds than others, but purely numerical comparisons can be made both on the basis of what happens on average, and what chance you have of winning a substantial prize.

Make the first comparison by seeing what happens, on average. For example, suppose you have £100 to spend in one of the three gambles. A week later (ignoring the dead period before your Bond qualifies for entry to the draw), on average you have about £25 from the Pools, £45 from the Lottery, and £100.10p from Bonds. No real contest there.

A better comparison for the average might be to look at three different ways of using a capital sum of £5000: choose between buying 5000 Premium Bonds, or placing the money on deposit to use the interest either to enter the Lottery, or to play the Pools. Buying Bonds gives an average return of 5% per year, which works out at £250. This average contains an almost invisible chance of winning a million pounds, and the most likely total you will get is £150 from three £50 prizes, but averages are averages: Bonds score £250.

It may be reasonable to assume that the capital, left on deposit, will generate about £5 each week for the Pools or the Lottery. A 25% payback on the Pools brings £65 over the year, on average. If you play the Lottery every draw, your average return is 45%, say £117, but you can push this up a little by confining your entries to those draws when the jackpot is boosted through a rollover or a superdraw. Other gamblers buy extra tickets at such times, taking advantage of this free money, but you might expect to push your average payout to 60% by this opportunistic strategy—say £156. However you do the sums, Premium Bonds are well ahead of the Lottery, with the Pools a distant third, on the basis of *average* return.

A fairer comparison for the Lottery or the Pools is to look at the prospect of winning some life-changing sum. To most people who buy Lottery tickets or fill in their Pools coupon, the average return is unimportant. The odds are remote, but zero participation = zero chance. Continuing with the three alternatives for this capital of £5000, look at a period of 20 years, and seek to compare your chances of *at least one win* during that period of either (1) £1 million or (2) £100 000 (chosen as benchmarks that might represent the ability to retire, or the ability to fulfil some lifetime dream). We use the figures that pertained to late 1997. As tastes change, you should be able to update the answers using more recent data.

Holding 5000 Premium Bonds over 20 years gives 1 200 000 separate chances to win the monthly prizes. There are some 8.4 billion Bonds, so your chance of ever winning one of the million pound prizes is about one in 7000. For the Lottery, historical figures show that about one ticket in 22.5 million wins a prize of at least one million pounds (and sometimes a lot more); over 20 years, your money will buy 5200 tickets, so your chance of ever being lucky is about one in 4300. The popularity of the Pools has been declining, and our calculations may be more favourable than the Pools deserve, but suppose the average of one jackpot of a million pounds each month holds good. Your money will buy about 1600 lines a month, among a total of about 2.5 billion lines entered. In any month, your jack-pot chance is the ratio of these numbers; over 20 years, you have about one chance in 6500 of hitting a jackpot. For the prospect of that elusive one million pounds, the Lottery is clearly best, with the Pools and the Bonds not very different from each other.

For the more modest £100K, we use the same methods. With Bonds, about six prizes per month are this amount or more, so your chance of ever winning that sum is about one in 1200. For the Lottery, about one ticket in 4 million has won at least £100K, so your chance over 20 years is about one in 800. For the Pools, assuming prizes of £100K or more are about six times as frequent as prizes of a million pounds, your chance over 20 years is about 1 in 1100.

For either of these two targets, and with £5000 capital, the Lottery is a clear winner, with little to choose between the other two. With other targets, and other capital sums, nominal interest rates, or periods of time, the answers will change. Doing your own sums is not difficult with remote chances. For your chosen target, and each possible gamble, estimate the chance, p, of success in *one* attempt. Count up the number of attempts, N, over the period of time chosen. So long as $N \times p$ is less than about 0.1, then the overall chance you will be lucky will be close to Np. Of course, you could just leave your capital on deposit. More certainty, less fun.

Postscript (November 1999)

To date, about one Lottery ticket in 20 million has won £1 million or more. The notional interest paid on Premium Bonds, and the chances of winning some prize, have both reduced. Despite attempts to revive Foot-ball Pools, the typical maximum jackpot has recently been about £600 000. Comparisons now favour the Lottery more than before.

CHAPTER 4

One coin, many games

It may be just a 50p piece to you, but to a probabilist a coin is also a convenient randomizing device. No-one uses paper money or credit cards to settle a family argument over which set of parents to visit at Christmas, or which TV programme shall be watched. Faced with two alternatives, you can allocate Heads to one of them, Tails to the other: if, when Tails appears, your instinct is 'Make it best of three', you know which one you really prefer. You do not have to obey the coin's whim, it might just help you make up your mind. Many recreational games are based round the easy availability of a coin which will yield two outcomes that can plausibly be taken as equally likely.

No coin is perfectly symmetrical, and so it is quite likely it has a small bias one way or the other. But, provided the results of successive tosses are effectively independent, we can use any coin to give two outcomes that will have exactly equal chances; label these as 'Heads' and 'Tails', and we have our fair coin. To accomplish this, simply toss the coin *twice*. If you get HH or TT, discard the results and toss twice more. Eventually you will get either HT or TH, and whatever the chance of H or T on a single toss, symmetry demands that HT and TH are equally likely on two independent tosses. So by labelling HT as Heads and TH as Tails, a 'coin' that gives Heads or Tails, each with a 50% chance, has been generated. This chapter is based round this ideal coin.

How's your intuition?

Just to get a feel for the range of possibilities, imagine six tosses of our ideal fair coin. The outcomes can be listed as

HHHHHH, HHHHHT, HHHHTH, ... ,TTTTTH, TTTTTT;

there are $2^6 = 64$ of them (Appendix I) and, on grounds of symmetry, all are equally likely. So the probability of getting all Heads is 1/64, the probability that all six tosses are alike is 2/64 = 1/32. To find the probability of equal numbers of Heads and Tails, we have to count how many of these outcomes have exactly three Heads. To do this, we count how many different ways there are to select the positions of the three Hs, as when these are fixed, the three Ts slot into the remaining places. Using Appendix I, there are $^6C_3 = 20$ ways of choosing where the Hs go, which means that the chance of equal numbers is 20/64 = 5/16.

Plainly, to have equal numbers of H and T in any sequence, the total number of tosses must be even. Let's dispose of one fallacious piece of folklore—that as you increase the number of tosses as 2, 4, 6, 8, 10, ... , so equal numbers of H and T become more likely (another wrong version of the Law of Averages). This is not just wrong, it is wrong in a spectacular way—the more tosses in this sequence, the *less* likely are equal numbers of H and T!

It is not hard to show this must be so. For any (even) number of tosses, we count up how many outcomes there are altogether, and how many of them have equal numbers of H and T. The chance of equality is the ratio of the second to the first—e.g. 20/64 = 5/16 as we saw for six tosses. Using the methods of Appendix I, the first few answers are shown in Table 4.1.To make comparisons easy, the bottom row of the table gives all the probabilities as fractions with a common denominator. It is easy to see that these probabilities decrease as the number of tosses increases—at least as far as is shown.

Spotting patterns is always helpful. In the change from two tosses to four tosses, the second answer (3/8) is just 3/4 of the first answer (1/2). Then as we move from four tosses to six tosses, the second answer (5/16) is just 5/6 of the first (3/8); similarly, going from six to eight tosses, the factor is 7/8. Here are these successive factors:

$$3/4 \quad 5/6 \quad 7/8.$$

3, 4, 5, 6, 7, 8—this must be more than a coincidence! With a leap of faith, we can 'predict' that as we go from eight to ten tosses, the chance of equality is reduced by a factor 9/10—and so it is, as you see from Table 4.1. It takes little courage to suggest that for 12 tosses, the chance of equal numbers of H and T will be 11/12 of the chance of equality with ten tosses. Indeed it is.

Table 4.1 For different numbers of tosses of a fair coin, the corresponding chances of equal numbers of Heads and Tails.

Number of tosses	2	4	6	8	10
Chance of equality	1/2	3/8	5/16	35/128	63/256
(Common denominator)	128/256	96/256	80/256	70/256	63/256

This pattern continues for ever, and the chance of equality steadily decreases. The next value is always found by multiplying the present probability by a predictable fraction that is less than one.

In 1730, James Stirling published one of my favourite pieces of mathematics, now known as Stirling's formula. This formula is to coin-tossing what an automatic dishwasher is to a dinner party—a welcome friend at a time of need. You want to know the chance of 50 H and 50 T in 100 tosses? Easy, with Stirling's formula—the answer is very close to $1/\sqrt{(50\pi)}$, about 8%.

This appearance of π looks weird. Our problem seems to have nothing to do with areas or circumferences of circles! But π is one of those numbers that pops up all over the place in mathematics, and this is one of them. For ten Heads and ten Tails in 20 tosses, the formula leads to $1/\sqrt{(10\pi)}$, which evaluates to 0.1784; the exact answer, at this level of accuracy, is 0.1762. Stirling's formula will not give a misleading answer. For future reference, the general result is

the chance of n H and n T in $2n$ tosses of a fair coin is very close to $1/\sqrt{n\pi}$

especially for large numbers of tosses. This approximation always exaggerates the chance of equal numbers by a small amount.

That result also shows something extra. If you toss a coin often enough, you can ensure that the chance of equality is as small as you like! You, madam, desire the chance to be under one in a hundred? No problem, with just a pocket calculator. Making $1/\sqrt{n\pi}$ less than 1/100 is the same as making $2n$, the number of tosses, more than 6366.2. Add a bit on to be on the safe side, so 6500 tosses will meet your request comfortably. A round number usually looks better as a working estimate.

And you, sir, want the chance of equality to be less than one in a thousand? 650 000 tosses will do it—so will 640 000, but you get my drift. Whatever challenge you give me (unless you unreasonably ask for a *zero* chance), I can meet it. For a million or more tosses, the chance of *exact* equality of H and T is tiny. But as you continue to toss, we know that the *proportion* of Heads gets close to 50%, and stays close.

Game on

Tossing a single fair coin against one opponent, the loser paying £1 to the winner, is a fair game: neither of you has an advantage. It is also rather dull. How could we inject some spice into coin-tossing?

One possibility is the St Petersburg game. You will toss the coin, and count how many tosses you need to throw H for the first time; this might take one toss, or it could take rather more. Very occasionally, it could take 20 or more tosses. The more tosses you need, the larger the prize you will pay to Suzy, your opponent. If you need only one toss, you will pay £2; if it takes two tosses, you pay £4; three tosses pays £8—you double the payout for each toss it takes. What is the catch?

The catch is that Suzy has to pay you a fee, up front, to enter the game. What would be a fair fee? How much would *you* pay, if the roles were reversed?

Averages usually help. If we can find the average amount she will win, she ought to be prepared to pay up to that. To work this out, seek the probabilities of winning the different amounts. Tossing stops when the first Head arises, so the possible outcomes are H, TH, TTH, TTTH, and so on. Appendix II shows their respective probabilities are 1/2, 1/4, 1/8, 1/16 etc., and the pattern that emerges is shown as Table 4.2. As the amounts she wins double, so the probabilities of obtaining them halve.

Now to find the average. Recall (Appendix III) that you multiply the amounts by the corresponding probabilities, and add them all up. Taking them one at a time, we begin with $2 \times (1/2) = 1$; the next is $4 \times (1/4) = 1$, followed by $8 \times (1/8) = 1$, and so on. So the total is:

$$£(1 + 1 + 1 + 1 + 1 + 1 + \ldots).$$

Hang on a minute! The sum never stops! Like Topsy, it just grows. Take enough terms, and it is more than £100; more than £1000; more than £1 million. You name it, this sum exceeds it. There is no finite value, so we reach the uncomfortable conclusion: *Suzy's average winnings are infinite.* However rich she is, Suzy ought to be prepared to hand over to you her entire fortune for the privilege of playing just once—on average, she wins an infinite amount, and so is ahead!

This conclusion is deeply suspicious, but there is no sleight of hand. The sums really do work out this way. But there is a flaw in the game's set-up: even if you are as rich as the Duke of Westminster (he owns most of

Table 4.2 In the St Petersburg game, the first few probabilities of winning the various prizes.

Amount won (£)	2	4	8	16	32	etc.
Probability		1/2	1/4	1/8	1/16	1/32	etc.

Mayfair—the real one), you only have a finite amount of money, so your hypothetical offer to pay sums in excess of £500 billion in certain circumstances is worthless. Once we cap the amount that will be paid, the calculation changes dramatically, and the average is no longer infinite.

For example, perhaps you will never pay more than £64. The game is the same, except that if the sequence begins with TTTTT, you will pay Suzy £64 only, however many more tosses you need to get the first H. The chances of winning up to £32 are as in Table 4.2, and the story is completed by paying £64 with the remaining probability, 1/32. The average Suzy wins is then

$$£(1 + 1 + 1 + 1 + 1 + 2) = £7.$$

With a ceiling of £64 on the payout, an entry fee of £7 is fair.

Suppose Suzy plays under these conditions. Three times in four, when H appears within two tosses, she makes a small net loss. Otherwise, she is up to £57 in profit. On average, losses balance gains, but prepare for a rollercoaster ride.

Suzy has to work out for herself what a fair entrance fee should be. If your scruples are no better than mine, you could offer to 'help' with the calculation. Point out that there are six possible prizes, £2, £4, ... , £64, and so the average she can win will be

$$£\frac{2 + 4 + 8 + 16 + 32 + 64}{6}$$

which works out as £21. Since you feel in a generous mood, instead of charging her the full £21 to play, you can do a special deal at just £15. Never give a sucker an even break. You'd never fall for my con trick, would you?

Leaving aside that piece of semi-plausible skulduggery, return to the full St Petersburg game. The problem arose because the average winnings were computed as infinite. If we change the payouts, that may no longer be the case. Instead of *doubling* the rewards each time, make them £(2, 4, 6, 8, 10, ...) according as 1, 2, 3, 4, 5, ... tosses are needed to produce a Head. This time, although the possible payouts still increase without bound, the correct calculation of the average yields a sensible answer: £4. (This is a nice exercise in A level maths.) So if Suzy pays £4 to enter, the game is fair to both sides—the average profit each play is zero. This change still asks you to offer an unlimited prize, so it is probably sensible to agree a reasonable maximum payout. As it happens, the sizes of the potential prizes increase so slowly that capping the payout at £64 makes no noticeable difference whatsoever—£4 remains the fair entrance fee. Were you to select £16 as the maximum prize, then the fair entrance fee reduces only to £3.98.

Penney-ante

This game, described by W. Penney in 1969, is a bear-trap for the unwary. It is easy to set it up in such a way as to make it appear to favour one player, when it actually strongly favours the other. So the potential for using your superior knowledge to increase your bank balance is considerable.

Take a fair coin, and toss it repeatedly. Here are the results of such an experiment with 30 tosses, that I carried out. For ease of reading only, they are grouped in blocks of five.

HTHTT   THTTT   HHTTH   HTHTT   HTHTH   HHHTT

Penney-ante is based on patterns of length three; the first pattern was HTH, and then THT arose at tosses (2, 3, 4); then HTT, TTT and so on. There are just eight different patterns to consider:

HHH   HHT   HTH   THH   HTT   THT   TTH   TTT

You allow your opponent, Doyle, to select one of these eight as 'his' pattern. You must choose from the seven that remain. After the choices have been made, one of you—or a neutral third person—begins a sequence of coin tosses. The winner is the person whose pattern appears first.

This game seems to favour Doyle. He has a full choice from all possible patterns, you have to select from those left, after he has presumably taken the best. At any rate, there appears no *disadvantage* in having first pick. Chess players will choose White, poker players prefer Aces, many Monopoly players want Mayfair and Park Lane. As you will have gathered by now, Doyle is being set up. ('Doyle', because the name of the victim in the classic 1973 film, *The Sting*, was Doyle Lonnegan.)

The essential fact is that, whatever Doyle chooses, you will be able to select a pattern that will appear before his choice at least 2/3 of the time. This looks an utterly incredible claim, on several accounts. Firstly, if you name any three tosses—numbers 6, 7, and 8 say—symmetry demands that all eight patterns have the same chance, so why should one pattern be likely to arise before another? And if one pattern *is* more likely to appear early, surely it is best to have first choice?

One way to see through this specious bluster is to imagine Doyle chooses HHH. You will then carefully select THH. I claim your chance of winning in these circumstances is 7/8. Why?

The one time in eight that the first three tosses are HHH, he wins immediately. The remaining 7/8 of the time tosses continue. Focus on what happens in this case. Sometime down the line, after the third toss, his pattern HHH will appear for the first time. In the example of the 30 tosses at the start of this section, it happened on tosses 25, 26, and 27. Now since this was the *first* time that HHH had arisen, it is impossible that toss 24 could have been H—otherwise HHH would have already appeared at tosses 24, 25, and 26! So toss 24 was inevitably a T. That means that the pattern THH certainly arose at tosses 24, 25, and 26, which is *before* the first HHH. (There is also the chance that our pattern THH occurred even earlier, well before this first HHH.) *Whenever HHH arose for the first time, THH had already happened.* So 7/8 of the time, THH appears before HHH. Unless Doyle wins at the first three tosses, he is doomed to lose.

I will not include the full arguments for all the choices Doyle might make: that would take some time, and most of the analyses are more complicated than the HHH v. THH. But Table 4.3 contains the relevant information for you to succeed. For each of the eight choices he might make, it shows your best choice, and your corresponding chances of winning. As claimed earlier, these chances are all 2/3, or even greater.

It would be foolish to let Doyle see you consulting your private Table 4.3, to ensure you choose the right response to his selection. Here is a simple mnemonic. Look at the *first* two members of his pattern and make them the *last* two in yours; and never choose a palindrome ('the first shall be last ... '). That will guarantee that you win at least 2/3 of the time!

Table 4.3 In Penney-ante, your best response to Doyle's choice, and the corresponding chance that your pattern appears first.

Doyle's choice	Your response	Chance of winning
HHH	THH	7/8
HHT	THH	3/4
HTH	HHT	2/3
THH	TTH	2/3
HTT	HHT	2/3
THT	TTH	2/3
TTH	HTT	3/4
TTT	HTT	7/8

You will naturally expect to play this game at even money. You could push your luck, and pay £5 when he wins but collect only £4 for your victories, but only make such an offer when he realizes he is at a dis-advantage. So long as you pay him less than twice what he pays you, you'll be ahead in the long run. (Being ahead over just half a dozen games is *not* assured.)

Table 4.3 throws up the nice cycle in Fig. 4.1.

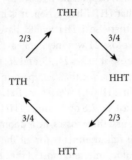

Fig. 4.1

Each pattern beats the next in sequence, in the sense that it is more likely to appear first! So, even among these four, there is no such thing as a 'best' pattern—whatever we select, another is superior.

Penney-ante can be played with patterns of any length, but three is a good choice. It usually does not take long for a winner to emerge, and whatever your opponent selects, you can always find a better pattern. This last point is not true with patterns of length two. If he chooses HH, you can smile and pick TH, winning 3/4 of the time. But if he selects TH, the best you can do is to pick either TT or HT, which give you and Doyle equal chances. With patterns of length four or more, you always have a winning strategy if you are allowed to choose second, but sometimes the games will be protracted.

During your games of Penney-ante, you might idly think about the length of time you have to wait for a pattern. Table 4.3 gives no informa-tion about this. One line of thought could be to argue as follows: 'The chance of any specified pattern starting at any particular position is 1/8, by symmetry. If an outcome has probability p, we know that the average time to wait for it is $1/p$ (Appendix III), so the average time to wait for a pattern—any pattern—is 8.' That argument would be perfectly valid if the outcome of interest depended only on one toss of the coin. But here we are looking at blocks of three tosses which overlap with each other, so the outcomes are not independent. It seems plausible that this should make a difference, but it is not easy to assess what sort of difference. If you play a lot of games, and make a note of the sequences, you may well get the impression that HHH and TTT do seem to take a long time to turn up.

That impression would be correct. Taking the eight patterns in the order shown for Doyle in Table 4.3, the respective average waits are:

$$14 \quad 8 \quad 10 \quad 8 \quad 8 \quad 10 \quad 8 \quad 14$$

tosses. The patterns HHH and TTT take nearly twice as long to appear as some of the others, on average. The other palindromes HTH and THT are a little lethargic too. The calculations leading to these answers are a bit complex, but if you want to try the challenge of a subtle, and powerful, argument, follow that given in the box below. There, we look at the

---

It takes *six* tosses to get HH, on average

The key is good notation. Write X to mean the actual number of tosses needed, so that X could be 2, 3, 4, 5, . . . —anything above 1. Whatever X is, it has some average that we will call A; this A is what we are calculating. Although X varies from experiment to experiment, A is a definite, as yet unknown, quantity.

When we embark on a series of tosses, half the time the first toss is a T and this gets us nowhere. In those series that begin with T, the average number of tosses to get HH is $1 + A$ (one wasted toss, and then, on average, A more). Otherwise, half the time we begin with a H; look then at the next toss, giving the first two tosses as HT and HH each 1/4 of the time. The sequence HT is a dead end—two wasted tosses, and we have to begin again; if we start with HT it takes, on average, a total of $2 + A$ to get HH. Finally, bingo! If we start HH, we have achieved our goal in two tosses.

Collect these three alternatives together:

- With probability 1/2, the average is $1 + A$.
- With probability 1/4, the average is $2 + A$.
- With probability 1/4, the number of tosses is 2.

But the *average* of these alternatives must also work out as A, i.e.

$$A = (1+A) \times \tfrac{1}{2} + (2+A) \times \tfrac{1}{4} + 2 \times \tfrac{1}{4}$$

which solves easily to give $A = 6$, as claimed.

simpler case of patterns of length two; it turns out that the average wait for each of HT or TH is four tosses, but for TT or HH the average wait is six. (The average time to wait for either T or H on its own is two tosses, shown by conventional calculations in Appendix III.)

The answers given for these average waits are unexpected. My natural instinct, when I first met them, was to disbelieve them. But intuition is fallible—if it were not, logic would have no place in human affairs. Like *Star Trek*'s Deanna Troi, we might be able to 'feel' our way to the correct conclusion, without necessarily explaining how. But sometimes, we need the logical approach of Mr Spock, or Commander Data. How might they, knowing that it takes an average of 14 tosses to get the first HHH, best explain why, to Kirk or Picard?

Spock, with his encyclopaedic knowledge of twentieth-century affairs, would draw Kirk's attention to the notorious number 11 bus: you wait with increasing frustration for half an hour, and then three come along in quick succession. So it is with the appearances of HHH—the key is *overlap*. Look again at the results of the 30 tosses I reported earlier. The triple HHH arose for the first time at tosses 25, 26 and 27. But once it had arisen, there was an immediate 50% chance it would happen again: and H at toss 28 did give HHH also at tosses 26, 27 and 28; further, if toss 29 had also been H, another HHH would have occurred. In other words, occurrences of HHH tend to be in bursts—just like number 11 buses.

The average wait between *all* appearances of HHH, in a long sequence of tosses, tends to arise from averaging one long gap with several short ones. But the time to wait for the *first* appearance of HHH is normally the initial long wait, before the burst. So the first appearance is not a typical one, and its wait is longer than average.

Would Kirk have the grace to accept Spock's analogy?

Notice that the four patterns {HHT, THH, HTT, and TTH} where the average wait is only eight tosses are precisely the patterns where the *last* toss (or tosses) is no help in getting that pattern again immediately. Patterns HTH and THT have an average wait of ten tosses: in each case, the last toss can be the first in the next appearance of the pattern, after only two further tosses. For HHH and TTT, with an average wait of 14, *both* the last two tosses help, and can lead to a repeat after one more toss only. The more helpful a pattern is to an immediate repeat of itself, so that patterns can more easily arise in bursts, the longer the average wait for the first appearance.

How long are you in the lead?

Let's return to the idea of a long sequence of tosses, and look at cumulative numbers of Heads and Tails. At any stage, either H leads, or T leads, or we have equality. Equality can only occur after an even number of tosses. How long might we have to wait for equality? How often will the lead change? Looking backwards from the end, when was the last time we had equality? The answers to such questions are very different from what most people expect.

It is plain that many games can be constructed around these questions. You can offer odds against the various outcomes, take in bets and run an experiment. You need a sound appreciation of what is likely, and what is unlikely, or you will be at the wrong end of the odds. At a non-mercenary level, you can have much entertainment from thinking about these questions, offering predictions, and seeing the answers. Take a time out now.

Be prepared to be surprised. William Feller, whose work inspired a generation of probabilists, was himself astonished. The first edition of his book, in 1950, was intended to be readable by non-mathematicians, so as to develop the reader's intuition about chance. The main change for the second edition (1957) was the inclusion of a chapter on coin-tossing. His preface includes 'The results concerning fluctuations in coin-tossing show that widely held beliefs ... are fallacious. These results are so amazing and so at variance with common intuition that even sophisticated colleagues doubted that coins actually misbehave as theory predicts.' Ten years later, he wrote a third edition, whose main changes were in this same chapter. There was nothing wrong in his previous material, but Feller had found a new and more enlightening approach, that avoided the use of complicated tricks.

If you have the mathematical capacity, read Feller's account. Here I will simply describe the answers to the questions posed, without a complete justification. Collecting your own data can help give a better feel for these ideas. On the principle that a pilot run can help, let's look at a sequence of just 12 tosses. The number of outcomes is small enough to be counted, and exact probabilities can be calculated.

How long to equality?

If you keep on tossing, you will return to equal numbers of H and T sometime, but not necessarily within 12 tosses. Table 4.4 shows the chances of a

Table 4.4 In a sequence of independent tosses of a fair coin, the respective chances that the numbers of H and T become equal again *for the first time*.

Number of tosses	2	4	6	8	10	12
Chance of first return	1/2	1/8	1/16	5/128	7/256	21/1024

first return at different times; there is a residual probability of 231/1024, about 23%, that the first return to equality is later than 12 tosses. Note the difference between Table 4.1 and Table 4.4: the former is concerned with *any* return to equality, Table 4.4 only to the *first* return.

If you do not equalize early on, it might well take a very long time. But when you do equalize, the subsequent path is independent of what happened earlier. That means that half the time you equalize again after two more tosses—and then perhaps another equalization follows in two further tosses. This is reminiscent of the number 11 bus analogy given above. Returns to equal numbers tend to cluster together, with longer gaps in between. But in this sequence of coin-tossing, there is now a nasty twist: the *average* time you have to wait for the first return to equality is *infinite*, so extremely long waits will certainly occur. (Sometimes, when waiting for the bus, you may well think life is imitating coin-tossing!)

How often will the lead change?

Let's be clear what we mean by a change of lead at a particular time. Two things must happen:

(1) At that time, the totals of H and of T are exactly equal.
(2) Whichever of H and T was in the lead at the previous toss, the other takes the lead at the next toss.

So the lead can change only at an even number of tosses, although we need to look one toss ahead. Because of this, we can eliminate any ambiguity by stopping to count the number of lead changes only after an *odd* number of tosses. That seems a sensible convention to adopt.

Before you read on, pause now to think how often the lead might change. Here are two specific problems: firstly, for a sequence of 101 tosses, on which of three alternatives would you place your money:

- four or fewer changes of lead?
- five to nine lead changes?
- ten or more lead changes?

Secondly, consider a sequence of 20 001 tosses. Make your guess as to the *most likely* number of lead changes over that whole period. To give you a clue, I offer the free information that the average number of lead changes in this long sequence is about 56, and you should do some lateral thinking: turn your attention to likely *scores*, and *averages*, of batsmen at cricket, before I reveal the answer.

About 20 cricketers have batting averages in Test matches of between 50 and 60 runs. Amongst these batsmen, in all their collective 2000 innings of Test cricket, what would you assess as the *most common* score in a single completed innings? The possible scores are 0, 1, 2, ..., with 375 from Brian Lara as the current maximum: but if you follow cricket closely enough, you will make the same guess as I do for the single most common score. Zero.

I have no intention of hunting through the record books to check my guess. But a few minutes with the individual scores in a dozen cricket matches will point to zero as being more frequent than any other score. And so it is with tossing a coin 20 001 times (indeed, any number of times): it is *more likely* that the lead between H and T *never* changes hands than any other specified number of changes. Further, if you look at the alternatives of 1, 2, 3, ... changes of lead, the probabilities *decrease* steadily: one change is more likely than two, two changes are more likely than three, and so on. (I would bet on the same phenomenon for cricket scores, up to a reasonable level, over the long term.)

All of the results I have just stated follow from one remarkable formula. The pity is that the proof of this formula is more mathematical than I care to include, so you will either have to simply accept what I say, or turn to somewhere like Feller's book to read it, or, if you a professional probabilist slumming it, work it out for yourself. The formula relates the number of changes of lead during the whole sequence to the total number of Heads in the experiment. That in itself is quite astonishing, as the former needs information on what happened at every single toss, whereas for the latter we require only the totals at the end.

> *The chance of exactly x lead changes is twice the probability that Heads is 2x + 1 ahead at the end.*

Take the example of 101 tosses. According to this formula, with $x = 0$, the chance of *no* changes of lead over the whole sequence is twice the probability that H leads by one at the end, i.e. twice the probability of 51 Heads and 50 Tails. Take $x = 1$; we see that the chance of *one* change of

Table 4.5 In a sequence of 101 tosses of a fair coin, the probabilities of the
numbers of times the lead between H and T changes hands.

Number of lead changes	0	1	2	3	4	5
Probability	0.158	0.152	0.140	0.125	0.107	0.088
Number of lead changes	6	7	8	9	10	11 or more
Probability	0.069	0.052	0.038	0.027	0.018	0.026

lead is twice the probability Heads leads by 3, i.e. of 52 H and 49 T. And the
chance of *two* lead changes is twice the probability of 53 H and 48 T, and so
on. The first few answers are shown in Table 4.5. These chances decrease as
I claimed, with the biggest probability at zero changes of lead. This
property applies to any number of tosses, not just to 101.

You can now check how you would have fared with your bet. Up to four
lead changes will occur 68% of the time, five to nine about 27%, and ten or
more a miserable 4–5%. In fact, in 31% of all such experiments, the lead
will not change more than once! Don't feel discouraged if your guess was
way too high—so was mine. The number of changes of lead tends to be far
fewer than most people suppose. You are likely to find many takers at even
money if you offered to bet on four or fewer lead changes in 101 tosses;
and you can expect to win such bets more than two times in three.

To be on the more favoured side in similar bets, you need to know the
break-even number of lead changes. For a sequence of N tosses, work out
the square root of N, and divide your answer by 3. You won't be far wrong.
To give yourself an edge, increase this by a bit—20% should do—and offer
to bet on *no more than* that number of lead changes. For example, when
$N = 101$, its square root is about 10, and dividing by 3 gives an answer of
just over 3; increasing this by 20% means you will be prepared to bet on
four or fewer lead changes—the case discussed above. For an experiment
with 251 tosses, the odds actually favour you if you bet on five or fewer
changes of lead. Almost unbelievable; and 10% of the time, there will be *no
lead changes at all*!

When was the *last* equality?

Over a sequence of tosses, sometimes Heads are ahead, sometimes Tails,
and sometimes they are level. If you carry on tossing a fair coin, inevitably
you will keep on returning to equality. In this problem, it is easier to think
about an even number of tosses, so take 200 as an example. If you look

back after 200 tosses, where in the history would you expect to find the *last* time H and T were level? It might have been at the beginning, when the score was 0–0; it must have been at one of 0, 2, 4, ... , 196, 198 or 200 tosses. You should not expect to be exactly right—there are 101 possible answers—but would you expect it to be relatively recently? In the middle? In the distant past? Choose your guess now.

The only way you have to go right back to the beginning is when there are no returns to equality during these 200 tosses. That seems very unlikely. Nevertheless, you should not be surprised if you have to go back quite a long way, because *half the time, the scores never equalize in the second half of the sequence!* The time the scores were last equal is just as likely to have been in tosses 0 to 100 as in tosses 100 to 200.

This means that, half the time, whichever of H and T is ahead at the halfway point stays ahead over the entire second half of the experiment. Imagine two friends taken hostage by a terrorist group, and allowed no books, chess sets or packs of cards to pass the time, just a single coin. They might heroically aid empirical probability by playing 5 hours a day, 600 tosses to the hour, every day over a year, and accumulating the scores. This is over a million tosses, yet this theory says that there is a 50% chance that the one who is leading at the end of June stays ahead during the whole of the rest of the year!

There are parallels with sport. In 50% of snooker matches between players of equal skill, the player ahead after 16 frames will remain ahead all the time until after frame 32. In soccer, will the team heading the Premier League at Christmas remain ahead for the rest of the season? (This question is more complicated, as this contest is between 20 teams, not just two.)

Working out the pattern of what to expect for any other number is easy, when you have seen the answers for 200 tosses. First, there is always total symmetry about the middle. This means that the chance that the time H and T were last equal was at 190 tosses is the same as the chance they were last equal at 10 tosses; the chance the last equality was at 2 tosses is the same as the chance it was at 198. (This curious fact was discovered completely by accident in 1964 by David Blackwell and two of his Californian colleagues. They were using their knowledge that two particular answers should always add up to 1 to check whether a computer programme was working correctly—always good practice. But the computer output unexpectedly showed a whole family of pairs of numbers that also added up to 1. Noting what these numbers corresponded to, they saw that the only possible explanation was this remarkable symmetry between 10 and 190, or 2 and 198, etc. Having realized that this result must surely hold for any number of tosses, they were able to prove this true in just 3 centimetres of mathematics!)

Because of this symmetry about the two halves of the sequence, once we know the probabilities for the first half, we automatically know them for the second. Aside from this symmetry, there is another useful feature: over the period of 0, 2, 4, ... , 98, 100 tosses, the associated probabilities *decrease*: so among these possible answers, the *most* likely time of last equality is back at zero, and the *least* likely time is after 100 tosses!

Bring these two statements together. If you want to bet on the single most likely time of last equality, you should select either 0 tosses, or 200 tosses—the chances are the same, and exceed any other individual chance. With either of these as your guess, the chance you are correct is about 5 or 6%. This is not particularly big, but it is the best you can do. Remember that there are 101 horses running in this race! If other people have chosen 0 and 200, and you must be different, your next best choice is either 2 or 198. The time of last equality falls in the small block of four times {0, 2, 198 and 200} in about 17% of all such experiments. By contrast, the *least* likely time of last equality is near the middle: lump together the 11 times {90, 92, 94, ... , 108, 110}, add up all the probabilities, and discover that the last equality falls in *this* block only a miserly 7% of the time.

Such answers astonished even William Feller, who knew a thing or two about fluctuations in tossing coins. Many people are inclined to plump for somewhere near the middle. This is as wrong as it is possible to be. It is quite practicable to set up a series of games with just 20 tosses, and ask opponents to guess when the last equality would occur. Table 4.6 shows the respective chances, so you can work out where your best bets lie.

The swift decrease of the chances as we move towards the middle is clear. In similar tables, for $2n$ tosses, the probability in the first box is well approximated by our old friend, $1/\sqrt{(\pi n)}$.

If intuition is such a poor guide when tossing coins, can we really trust it in more complex situations?

Table 4.6 Tossing a fair coin 20 times. The chances that the *last* equality between H and T was at different times are shown.

Time of last equality	0 or 20	2 or 18	4 or 16	6 or 14	8 or 12	10
Probability	0.1762	0.0927	0.0736	0.0655	0.0617	0.0606

Postscript

When two long queues stretch ahead of us—in the Post Office, a super-market, or in a traffic jam on a two-lane carriageway—isn't it frustrating how often we seem to make the wrong choice? Whichever queue we join, the other one proceeds faster. Some parallels can be drawn with the way the lead fluctuates in long sequences of tossing fair coins.

Consider the traffic jam. As you approach, the two queues look to have equal length. You make your decision which one to join, and have to stick with it, as other cars join both queues. You identify as your competitor a blue BMW that stopped adjacent to you; at some time in the future, are you ahead, level or behind? This is decided by some deity who begins a series of coin-tosses to select which of the two queues moves next. Our theory shows that long periods in which one or other of us remains ahead is the norm, and frequent lead changes is the exception. Random chance picks us or our competitor as the fortunate one, but memories are selective: we may remember the slow journeys, and not the ones where we guessed the faster queue. Or, having drawn well ahead of the BMW, we select a new benchmark, a green Mini, to measure our progress. The more battles we set up in this way, the more likely it is that the last one is lost.

Averages are little help. If the BMW moves first, how long do you wait to catch up, on average? Our coin-tossing work gives the depressing answer: an infinite time! Is this the reason we are unhappy with our progress? Take one consolation: the BMW driver does the same sums, and feels just as frustrated.

Summary

The games we have looked at have been based on tossing a coin that falls Heads or Tails, exactly half the time. However biased our actual coin, we have seen a way to achieve this ideal coin.

The St Petersburg game, like its variants, is centred on how often you have to toss to get the first Head. It is unsatisfactory in its original version, but if we cap the payouts, or do not allow them to increase too swiftly, fair games can be set up. You need to be able to work out averages, to establish a fair entry fee.

Penney-ante, waiting for particular patterns of length three, has a winning strategy for the second player. Whatever pattern the first player picks, the second player can always choose a pattern that appears first, at least two times in three. The 'better' patterns also tend to have shorter average waiting times to their first appearance. To assess the average time for a given pattern to arise, the key consideration is the amount of help given to its *next* appearance, should it arise now. The more helpful a pattern is to obtaining an early repeat of itself, the *longer* the average wait for its first appearance.

For a long sequence of tosses, look at the cumulative difference in score between H and T. Although the average difference at any time is zero, an actual difference of zero arises far less frequently than most people expect. Far, far less. It is more likely there will be *no* changes of lead in a sequence of any length than any other number of changes. Typically there are periods of enormous length with no equalization between H and T, and then perhaps several equalities near together. Looking back from the end, the last time H and T were equal is much more likely to have been right at the beginning, or right at the end, than over a whole raft of places near the middle. Untutored intuition seems as misleading in these problems as it is possible to be.

Test yourself quiz

(All the coins here are supposed *fair*.)

4.1 Use a pocket calculator with the result from Stirling's formula (p. 51) to find the probabilities of exactly equal numbers of Heads and Tails in (i) 50 tosses (ii) 250 tosses (iii) 1000 tosses.

4.2 Imagine playing Penney-ante, but with patterns of length four. Your opponent chooses HHHH; what would be your best reply? With what chance would you win? Taking the hint near Table 4.3, what might you choose if he selected TTHH?

4.3 Both Tables 4.1 and 4.4 are concerned with the chances of equal numbers of Heads and Tails. The difference is that the latter is about the chance of them being equal *for the first time*. Do a piece of arithmetic: for each given number of tosses, find the *ratio* of the probabilities in the two tables.

4.4 Accepting the assurance that the pattern you (should have) found in Q4.3 continues for ever, use your answers to Q4.1 to find the chances of Heads and Tails being equal *for the first time* after (i) 50 tosses (ii) 250 tosses (iii) 1000 tosses.

4.5 Consider a possible sequence of 1001 tosses: what is the single most likely number of times the lead changes hands? If you were allowed just one guess as to the last time Heads and Tails had appeared equally often, what would you choose?

4.6 In this same sequence of 1001 tosses, take the four statements about the number of changes of lead:

A. There will be at least 30.

B. There will be from 20 to 29.

C. There will be from 10 to 19.

D. There will be at most nine.

Without attempting to work their chances out, see if you can place them in order, with the most likely first and the least likely last.

Do the same thing for 10 001 tosses.

CHAPTER 5

Dice

Dice have been used in gaming and recreation for thousands of years. They were used in board games in Egypt over 5000 years ago, and in various games of chance invented during the Trojan Wars to relieve the boredom of the soldiers. The Roman emperors played: Augustus wrote to Tiberius about his experiences, and Claudius had a board fitted to his carriage so that he could continue to play while driving. In *Games, Gods and Gambling*, F. N. David notes that in hoofed animals, the astralagus, or ankle bone, is fairly symmetrical, because of the way a foot takes up the strain placed on it. Dice from ancient times were often constructed from this solid and robust bone.

It has been known for 2500 years that there are just five completely symmetrical solid shapes, the so-called Platonic solids. Perfect dice can have no other shapes. Aside from the familiar cubical die, these solids are:

- the tetrahedron, with four triangular faces
- the octahedron (diamond) with eight triangular faces
- the dodecahedron, with 12 faces, each a regular pentagon
- the icosahedron, with 20 triangular sides.

Dice with any of these shapes can be bought, but none has seriously rivalled the cube as the choice for gaming. Some early dice, with 14, 18, or 19 sides have been uncovered, but it is impossible for them to have all their faces exactly the same shape and size. Unless we say otherwise, our dice are taken to be cubes whose faces are numbered 1 to 6, all equally likely, and giving independent results each throw.

One die

How would we judge, statistically, whether a die is satisfactory? Just as in a similar question about the National Lottery, the answer depends on how large a difference from absolute fairness you wish to detect, and how much data you will accumulate. Perhaps you are prepared to throw the die 120 times, so that each face is expected to show 20 times? The statistical question is how much *variation* about that average is acceptable. As an illustration, suppose our die has been so badly constructed that three faces each have the below average probability, 0.147, while the other three faces each arise with probability 0.187. Would we expect to detect such differences in this experiment?

To answer this, turn to computer simulation. We can programme a computer to give whatever probabilities we like to each face, and explore what happens. The frequencies from four separate runs of the same programme, which mimics the sub-standard die just described, are shown in Table 5.1. Faces 1, 2, and 3 had been allocated the lower probability.

Given just *one* of these rows, representing the data from the experiment of 120 throws, you would find it impossible to pick up the true nature of this biased die, *even if you were told that exactly three faces had a below-average probability*. In the first row, you would surely bank on Face 5 being one of these—and you would be wrong. Row two shows exactly the sort of random variation to be expected in a completely fair die. In row three, you would correctly pick out Faces 2 and 3, but you have no sensible reason to select Face 1 as having below average probability. Although row four points to Face 1 as a rogue, none of the other five faces can confidently be separated out.

The lesson to be drawn is that only through a very long experiment can you have any confidence in being able to pick out anything except a gross bias. You need far more than 120 throws. Even adding the four rows together, which represents a single experiment of 480 throws, the favoured Face 5 showed slightly *less often* than the average for a fair die (79, as opposed to 80). Appendix IV gives more details on the analysis of these data.

For an idea of how much data you might have to collect, turn to the

Table 5.1

Face 1	Face 2	Face 3	Face 4	Face 5	Face 6	Total
19	21	18	20	13	29	120
15	20	16	26	22	21	120
25	14	11	22	21	27	120
12	21	24	21	23	19	120

experiments conducted a hundred years ago by the statistician W. F. R. Weldon. Details can be found in the 11th edition of the *Encyclopaedia Britannica*, in the article explaining Probability. Weldon's data included 26 306 throws of sets of 12 dice, which indicated clear evidence of a small bias, with Faces 5 and 6 appearing more often than could reasonably be put down to chance. There is no *a priori* reason to suspect your own dice exhibit these same faults, but you will realize how much work could be needed to detect, with any certainty, a small but significant bias.

Even in so simple a game as Ludo, it helps to realize how likely certain events are. You have just thrown a Six, and are deciding whether to advance a running man or start a new one. If you opt to run, what are the chances that you have to wait at least a dozen, or at least 20, throws for the next Six?

On average, you have to wait six throws for any particular face. The only way it takes at least a dozen throws for the next Six is if the next 11 throws are all non-sixes, and that (using independence of throws) has probability $(5/6)^{11}$, or about 13.4%. A similar calculation shows you have to wait at least 20 throws about 3% of the time. So if your running men are close to home, you are in some danger of wasting throws while waiting for a Six, unless you start a new man now. The independence of throws adds to your mounting frustration: it may seem like donkey's years since your last Six, but, on average, you still have to wait another six throws for the next one.

Ludo contains many positions where an appreciation of the chances will aid your cause. In the main, they are simplified versions of a parallel conundrum in that excellent game of sophisticated Ludo, known as backgammon.

Two Dice—backgammon

This is not the place to explain the basic rules of this game. Either you will know them, or you should find a friend to teach you. At every stage of the game, it is helpful to have a good appreciation of how likely are the possible rolls of two dice.

Modern backgammon is based on similar games played 5000 years ago, but was transformed in 1925 by the introduction of the Doubling Cube. At any stage, either player can offer to double the stakes; the other then decides whether to accept this, and play at doubled stakes ('take'), or to resign and concede the game to the doubler at the initial stake ('drop'). If play continues, the player who accepted the double now 'owns' the Cube, and he alone has the right to offer the next double, should the game move in his favour. Doubling and redoubling can continue in this fashion indefinitely (or with a pre-assigned limit), ownership of the Cube changing hands each time.

If doubled, under what circumstances should you take? To decide this, we should work out what happens, on average, for either taking or dropping, and go for the better option. Ignore for the moment any possible benefit from owning the Cube if you take.

If the current level of stakes is one unit, you lose one unit if you drop, but you play on in a game for two units if you take. There is some probability, call it x, that you will win if the game continues. If you take, you will win two units with probability x, or lose two units with probability $1 - x$; the *average* amount you win is thus:

$$2 \times x + (-2) \times (1 - x) = 4x - 2.$$

Compare this with the loss of one unit for dropping; the crossover point between the two decisions is when $x = 1/4$. *If your chance of winning is less than one in four, you should drop, if it exceeds one in four you should take.* This applies no matter how often stakes have been doubled.

We disregarded the advantage of owning the Cube if you take. If you take in the early stages of a game, you may have some lucky rolls, and see the game move in your favour. You can then put pressure on your opponent by redoubling. This advantage is more difficult to quantify, but indicates that it could be worth taking if your winning chance is less than one in four—but not when you face the risk of losing a gammon at double stakes!

A much more difficult question is whether or not to double. Doubling when your chance of winning is under 50% is plainly a bad idea, but suppose you can see you have the advantage. In this case, doubling doubles your average winnings—good—but you cede control of the Cube to your opponent. The next double can only come from him, and if the game moves in his favour, he will redouble with relish. So early in the game, even if you know you are more likely to win, you should be disinclined to double unless your advantage is substantial. Later, a double when you have only a slight advantage may pay off (but see Example B below). Should you have a substantial lead, and the chance of a gammon, think carefully before doubling: your opponent may happily accept the loss of a single unit, expecting a larger loss if the game had continued to the end. Despite these strictures, backgammon is a game of risk. If your opponents nearly always drop your doubles, and you never lose a game in which you have doubled, *you are not doubling often enough.*

Test your feel for the probability calculations that arise by seeking to solve the problems offered below. None of them are original, of course, and it is a pleasure to acknowledge Edward Packel's *The Mathematics of Games and Gambling* as a source of some of them.

Example A

The game is nearly over. It is your throw, you have men on your Five-point and your Two-point; your opponent has just one man on his Three-point, and so will finish next time (Figure 5.1). You hold the Cube. Should you double?

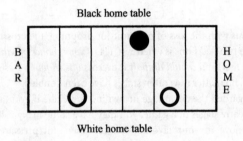

Black home table

B
A
R

H
O
M
E

White home table

Fig. 5.1

We solve such problems by counting. There are 36 possible results when you throw the two dice; 17 of these fail to win (all 11 that include a 1, also the 3–2, 2–3, 4–2, 2–4, 4–3, and 3–4). So your chance of winning is 19/36. His chance is 17/36, well above the crossover value of one-quarter, so he will take any double. If you do not double, you win, on average,

$$(1) \times (19/36) + (-1) \times (17/36) = 1/18$$

and twice that if you double. You should double.

Example B

This is exactly the same as above, except that your opponent's man is on his Five-point (Figure 5.2). What now? Should you still double?

Although your chance of winning is better than before (you might win

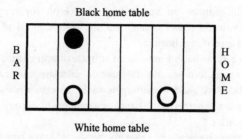

Black home table

White home table

Fig. 5.2

immediately, as above, or even on your second throw if he fails), this time you should *not* double! Here are the alternatives, if the game continues:

(1) You win immediately.

(2) You throw and fail to win, then he wins on his first throw.

(3) You get a second throw, as both of you fail to win on your first throws.

Take the initial stake as one unit, and suppose first that you do *not* double. Alternative (1) occurs with probability 19/36 as before, and you win one unit. But if you do not win immediately, it is now your opponent's turn, and there are just five throws (1–1, 2–1, 1–2, 3–1, or 1–3) where he does not win. His chance of winning is 31/36, which means you lose one unit with probability $(17/36) \times (31/36) = 527/1296$. You will get to stage (3) with the remaining probability, 85/1296, and if you do so you will definitely double. At that stage, your winning chance is so high that he will certainly drop, so you win one unit. Your average winnings are:

$$(1) \times (19/36) + (-1) \times (527/1296) + (1) \times (85/1296) = 121/648.$$

But suppose you double and he accepts. This is almost identical to Example A. With probability 19/36, you win two units immediately, but with probability 17/36 you fail to bear off, and *he controls the Cube*. It is now his throw, and he will redouble in a flash. His chance of victory, if we reach this point, is at least 31/36, so you should drop. Your average winnings if you double are:

$$(2) \times (19/36) + (-2) \times (17/36) = 1/9.$$

Since this is worse than before, you are better off *not* doubling.

Contrast these two examples. In the first one, it was best to double, but not in the second one, *even though you were better placed*. Distinctions such as these should caution you against lazy reliance on any simple rule to 'double if you are ahead', even near the end. Control of the Cube outweighed everything else here.

It plainly helps to be able to make an estimate of your chance of winning at various stages, but you do not have to calculate this chance to six decimal places! A good beginning is to be able to make reasonable judgements as to whether your chance of winning is under one-quarter, or over one-quarter.

Backgammon is essentially a race, so you should know how far you and your opponent are from finishing. When the game starts, your 'pip count' (the sum of all the distances your 15 men must move to bear off) is 167. The average number of pips you move in one throw is just over eight (as doubles double up), but the *range* of distances is zero to 24. A small number of lucky or unlucky throws can move you from well ahead to significantly behind in the race.

In a pure race, after there is no possibility of further contact, how much does it matter exactly what moves you make? Sometimes it makes no difference at all, but a serious player seeks every advantage, no matter how minuscule. It will pay off in the long run. Recall the 1974 soccer World Cup: progress to the next stage might depend on which teams beat Zaire most heavily. Yugoslavia attacked strongly even when the game was comfortably theirs, won 9–0, and were duly rewarded. Scotland had been content to win 2–0, and went no further.

Example C

You have three men left, on your Six-point, Four-point and One-point (Figure 5.3). You roll a Double One. You can remove only one man, but ought you to leave (6, 1), (5, 2), or (4, 3) to give the best chance of finishing next throw?

To find out, just count up how many rolls out of the 36 possible that will

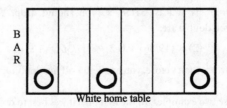

White home table

Fig. 5.3.

lead to finishing next time. In the order given, the answers are 15, 19, and 17, so you should leave (5, 2) to maximize your chance.

This sort of argument is easy to construct at the table, and you should take the time to do so. There are specific plays in the end game that a good player commits to memory, in case he ever has to use them. The more casual player seeks general rules that will be easy to remember, and will nearly always be best, or almost best. One such rule is to bear off as many men as possible. Another is to spread your men over as many different points in your Home Table as you can, so as to increase the chance that you bear off a man whatever the result of the roll. Generally these are good guidelines, but Example D illustrates where they can lead you to the wrong decision.

Example D

You have four men left; two are on your Three-point, the others are on your Six-point and your Five-point (Figure 5.4). The bad news is you throw 2–1, and you must maximize your chance of finishing *next* time.

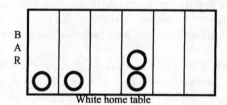

Fig. 5.4.

If you bear one man off from your Three-point, the only throws that work next time are 6–6 and 5–5. If you double up two men on to your Four-point—breaking both the rules about spreading them evenly *and* bearing off the largest number—*you also have an extra chance*, as 4–4 will also win for you. The increase is only from 2/36 to 3/36, but contemplate the warm glow of satisfaction when 4–4 appears!

There were special circumstances in Example D that led to breaking the rule about bearing off as many as possible. The choice was to leave three men, or four men, and in either case, you have to throw some double to bear them all off at once. If we had a choice of leaving two men or three men, it will *always* be best to bear off one man, and leave just two men. This distinction between having an odd number or an even number to bear off can make a large difference in deciding about doubling.

Suppose White has six men, Black has five men, and the pip counts are equal. If it is now White to throw, White is likely to be favoured, as both sides will probably need three throws. But in a similar position, if White had five men and Black had four, Black is likely to be favoured, as he may well need only two throws to White's three.

Example E

Consider these five positions near the end of the game, with a pip count of ten in each case. The main difference is in the number of men.

(1) Two men, at positions {6, 4}

(2) Three men, at positions {5, 4, 1}

(3) Three men, at {6, 2, 2}

(4) Four men, at {4, 3, 2, 1}

(5) Five men, at {3, 2, 2, 2, 1}

Let's compare their relative chances of finishing in one move, or in at most two moves. Before you read on, make as good a guess as you can about their chances; at least try to put these chances in the correct order, from best to worst.

It ought to be fairly clear that Position 1 is best, and Position 5 is worst, but assessing the other three is not at all obvious. To find the exact answer, you have to go through the possibilities systematically. I find the best way in such a problem is a two-step process: first, see what happens when the first roll is a double (six alternatives), then look at the other 15 possibilities for the first throw, remembering each can arise in two ways. For those first rolls that do not end the game, you may have to choose between alternative first moves, to maximize your chance of completion next time.

I will spare you the fiddly detail. Of the 36 possible first rolls, the numbers of outcomes that lead to bearing all the men off at once are, in the order of the five positions listed, {8, 3, 4, 3 and 0}. This demonstrates the very significant advantage of two men over three, and four men over five. There are $36^2 = 1296$ outcomes for two rolls (if needed), and when the dust has settled, the numbers of these that lead to completion within two rolls are {1203, 1187, 1047, 1127 and 335}. So the order, from best to worst, is 1, 2, 4, 3, 5, if your interest is in finishing within at most two moves.

The comparison may be easier in percentage terms. You will need only one roll from (6, 4) 22% of the time, and your chance of finishing in at most two rolls is 93%. From (5, 4, 1), the chance that one roll suffices is 8%, and you complete within two rolls 92% of the time. With your men at (4, 3, 2, 1), the chance you bear them all off the first roll is also 8%, but you have only an 87% chance of ending within two rolls. It may be surprising to find this better than (6, 2, 2), but here the chance of bearing all three off within two rolls is down to 82% (although you do have an 11% chance of success at the first roll). It is impossible to remove all five men at (3, 2, 2, 2, 1) with one roll, and you will complete within two rolls only 26% of the time.

A frequent decision in backgammon concerns leaving blots. What should you do when every legal move leaves some blot? What you need to know is the probability a blot can be hit, given how far away it is from an opponent's man. Sometimes you will hold intermediate points that prevent your opponent from hitting your man, even when he has the correct total; for simplicity, ignore this point—it does not affect the thrust of the main conclusion.

Take the example when your man will be five steps away from the enemy; you are vulnerable to any of the 11 throws that include a Five, and to the four other throws 4–1, 3–2, 2–3, and 1–4, so the chance of being hit is 15/36. When your man is eight steps away, you are vulnerable only to the six combinations 2–2, 4–4, 6–2, 5–3, 3–5, and 2–6, so the chance here is 6/36. To make complete comparisons, Table 5.2 shows the number of dice

Table 5.2 How many combinations your opponent can throw, and hit your blot, according to how far from it his nearest man sits.

Distance to blot	Number of combinations that hit
1	11
2	12
3	14
4 or 5	15
6	17
7 or 8	6
9	5
10 or 12	3
11	2
15, 16, 18, 20 or 24	1

Table 5.3 The average numbers of throws to bring on one or two men from the bar, according to how many Home Table points your opponent holds.

	One held	Two held	Three held	Four held	Five held
One man	1.03	1.25	1.33	1.8	3.29
Two men	1.32	1.69	2.22	3.24	6.25

throws that will hit a blot at a given distance. To find the probabilities, divide by 36.

Notice the discontinuity between distances six and seven. It is *much* easier to hit blots within six points than further away. And if you must leave a blot within six points, place it as close as possible to the enemy. Table 5.2 is also useful if you *do* want to leave blots, i.e. you wish to be hit so that you can re-enter on your opponent's Home Table and harass him.

When you have been hit, how many throws might you take to re-enter? Table 5.3 shows the average number of throws to re-enter one or two men, according to how many points your opponent holds. To see how the numbers are calculated, take the example when he holds four points in his Home Table. (This is another calculation where we first find the probability that an event does *not* happen.)

With one man on the bar, the only way you do not re-enter is when both your dice show one of the four numbers he holds. For either die, the chance of this is $4/6$ and so, using the independence of the dice, the chances you are unlucky with both are $(4/6)^2 = 4/9$. That means the chance you do re-enter is $1 - 4/9 = 5/9$, and so, on average, you will take $9/5 = 1.8$ turns to re-enter. (Appendix III explains this calculation.)

But suppose you have two men on the bar: you will only bring both back on to the board with your first throw if you are lucky with both your dice, which has probability $(2/6)^2 = 1/9$. Altogether, you have one chance in nine of bringing both on, four chances in nine of bringing on one man, and four chances in nine of wasting both dice. When all the calculations are over, you will need $81/25 = 3.24$ turns on average to bring both men on and start racing round the board again.

There are several lessons to draw from Table 5.2. Notice how much longer it takes to re-enter when four or more points are held, and also when two men are on the bar instead of one. For example, suppose you hold four points in your Home Table, and can put two opposing men on the bar, *provided* you leave a blot at your own Five-point. The risk may be well worth taking (depending on the positions of the other men): he is likely to waste several rolls, and more than two times in three you will have the chance to convert your blot to a point before it is hit.

Calculating or assessing the odds is needed at many stages in Backgammon. Good books will have based their advice both on their author's own experiences at the game, and on calculations similar to those offered here. There are now computer programmes capable of rolling out given positions thousands of times a second, using brute force to find the best play, when calculations are impracticable. Gary Kasparov has lost at chess to a computer programme; if silicon does not already rule in backgammon, its time is nigh.

Two dice—Monopoly

Since Clarence Darrow persuaded Parkers to market Monopoly (see Figure 5.5), versions have appeared in many languages, with local names for the locations. To aid international readers, we note the colours of the sets of properties as well as some UK names. The winning recipe is to acquire property, build houses or hotels, and rake in the rents to drive the others to bankruptcy. But with four or more players, you are unlikely to be able to collect a complete set without having to trade, so you need to know *which* sets are most desirable.

The four factors relevant to any particular set are:

- the cost of acquiring it;
- the cost of houses and hotels;
- the rent you charge when you have several houses or an hotel;
- its position on the board.

The first two points are obvious. The third is important so that you can assess how quickly you can bankrupt opponents (Mayfair in the Deep Blue set can charge up to £2000, usually a mortal blow; Old Kent Road in the Brown set never garners more than £250). The last factor, position on the board, is much more important than your casual opponents realize.

There are 40 squares on the board, and if players just toured round according to the score on the dice, all squares would be visited equally often in the long run. But there are jumps from one square to another, the principal example being the 'Go to Jail' instruction. *The Jail square is occupied rather more often than any other square.* How best to take advantage of this?

Fig. 5.5 Monopoly board.

The distance of each move is the total score on two dice, which ranges from two to 12. Scores in the middle are much more likely. Counting all the possibilities, the total is seven with probability 6/36, totals of six or eight each have probability 5/36, five and nine arise with probability 4/36 and so on, with two and 12 each turning up just once in 36 throws. So the Jail square is in a splendid position from which to hit one of the three *Orange* properties, situated six, eight, and nine steps away. The chance of an immediate hit is 14/36, and unless the escape from Jail is by a score of ten or more, there is a second chance to hit Orange. In contrast, a token leaving Jail hits the adjacent Purple set at one, three and four steps away less than one time in seven.

Three successive doubles also sends you to Jail, and some of the Chance or Community Chest cards direct players to specific places. A large computer simulation of a game, simply obeying the jump rules, can show how often each square will be hit. In such an experiment, representing ten million hits, the Jail square was indeed visited nearly twice as often as any other square. This experiment also showed how often the different property sets will be visited.

There are six sets that contain three properties. For every 100 hits on the Purple or Light Blue sets, there are about 110 hits on Green or Yellow, and 122 visits to Orange or Red. The Deep Blue and Brown sets have only two properties, so are hit much less often; for a fairer comparison, for every 100 hits on an 'average' Purple property, we find about 102 to an average Brown or Deep Blue. At the extremes, for every two visits to Park Lane in the Deep Blue set, there are three hits on Trafalgar Square in the Red set.

Inspect the Title Deeds to see how rent levels increase as more houses are built. The increase in rent in the change from two to three houses is *always* bigger than any other change. The marginal rate of return for building a third house is enormous—even loan sharks might recoil from demanding such a sum! Moreover, the rent chargeable for just three houses is often large enough to embarrass opponents. With two sets of properties, it usually makes better sense to have three houses everywhere, than hotels on one set and two houses on the other.

Each set of properties has its own basic economics. Find the total needed to buy the whole set, and place hotels everywhere—call this the 'Cost'. Add up all the maximal rentals payable—call this the 'Income'. Calculate the ratio of Income to Cost—the larger the better. This particular ratio has no direct meaning in the game, it is simply a way of making a comparison across the eight sets of properties. In descending order, with the value of the ratio in parentheses, we find Light Blue (1.59), Orange (1.41), Deep Blue (1.27), Purple (1.24), Yellow (1.15), Brown (1.13), Red (1.09), Green (1.01).

A different comparison is to ask how little you need spend over a set of properties, before every rental is at least £750 (rather arbitrary, I concede, but chosen as a sum that is likely to be large enough to cause your opponents a problem). This target is not achievable by Brown or Light Blue. For the other six, with the outlay needed in parentheses, the order is Orange (£1760), Purple (£1940), Deep Blue (£1950), Yellow (£2150), Red (£2330), Green (£2720). The value of the Deep Blue set is even better than these figures suggest, because the rents are well above the threshold of £750; for similar reasons, the Green set can be upgraded.

Collect together the pointers from the frequency of visits, the basic economics, and the investment needed to hit an opponent hard, and what emerges? At the bottom sit Brown and Green: neither attract large numbers of visits, Brown packs little punch, and Green is very expensive. Comfortably out in front is Orange: it is hit quite often, site and property are relatively cheap, and you have a good prospect of collecting damaging rents. The other five sets bunch together, deficiencies balancing attractions. This ranking should guide your trading when no-one has a complete set. (Be sure to demand some premium when haggling to swop an unwanted Green property for the coveted Orange one to complete the set. You did pay more to acquire the Green, and you do not want your opponents to realize how desperate you are for Vine Street.) Finally, think of placing four houses, rather than one hotel, on as many of your properties as you can: the number of houses in a Monopoly set is fixed, and if you hog them, you deny them to your opponents.

Modestly attribute your victories to pure chance. Otherwise your winning tactics might get noticed—and used back at you!

Two dice—craps

There are many variants on the basic game, depending on the side bets that are allowed. Like many dice games, its roots are very old, but its modern popularity is traced to its use as a pastime among soldiers in the First World War. You roll two dice and find their total. If this is seven or eleven—a natural—you win, but if it is two, three or twelve—craps—you lose. Otherwise the game continues with a series of further rolls, until either your first total is repeated (you win) or seven is rolled (you lose).

Plainly, the chance that the first total is seven or eleven is $6/36 + 2/36 = 8/36$. If you neither win nor lose immediately, you start on a possibly long series of throws to see which of two events occurs first. It looks as though computing the probability is complicated, as we do not know how many throws will be needed. But one simple idea removes this obstacle: if events A and B are disjoint, so they cannot occur simultaneously, and their respective chances are a and b, then the chance that A comes first is $a/(a + b)$. For example, suppose A corresponds to a total of six, and B to a total of seven. We know that A has probability $5/36$ and B has probability $6/36$, hence the chance that A comes first is $5/(5+6) = 5/11$.

To see why that is so, note that if neither A nor B occurs, we can simply ignore that throw. The only throws that are relevant are those that result in either A or B. Because their respective probabilities are a and b, in a long series of N throws A and B occur about Na and Nb times. Concentrating only on those throws that give either A or B, the occurrences of A and B are mixed up at random; the chance that A comes first is just the proportion of As in this list—and that is $Na/(Na + Nb) = a/(a + b)$.

We can now find the total winning chance by adding up the chances for each different way of winning. For example, the probability that your original total is six, and then that subsequently a six is thrown before a seven, is

$$\frac{5}{36} \times \frac{5}{11}.$$

Making similar calculations for initial totals of 4, 5, 8, 9, and 10, and not forgetting the 8/36 from winning on the first throw, the overall winning chance is 244/495, about 0.493. The game is slightly unfavourable to the player—who has the fun of rolling the dice.

In UK casinos, about 1% of total stakes are placed on this game; it is much more popular in the United States where gambling is not centrally regulated, and local rules and bets vary. But like all casino games, except possibly blackjack (see Chapter 10) you can expect to lose.

Three or more dice—totals

In some dice games, the aim is to get an exact total. How best to work out the respective probabilities, to see whether you are favoured? As often, it comes down to *counting*. But you have to know *what* to count. Take the example of three dice, seeking a total of nine or ten.

One argument goes as follows. Listing the ways in which these totals can arise, we find:

$$9 = (6, 2, 1), (5, 3, 1), (5, 2, 2), (4, 4, 1), (4, 3, 2), \text{ and } (3, 3, 3)$$

$$10 = (6, 3, 1), (6, 2, 2), (5, 4, 1), (5, 3, 2), (4, 4, 2), \text{ and } (4, 3, 3)$$

Since there are six ways of getting either nine or ten, the chances are equal.

If you believe that argument, you will believe almost anything! If *that* logic holds, then so does the following: 'the only way to total three is to throw (1, 1, 1), the only way to total four is via (2, 1, 1), so totals of three and four are equally likely.'

This argument, like the previous one, is false. The arithmetic in both instances is fine, but the whole truth has not been described. The fallacy in the argument can be seen by labelling the three dice as red, white, and blue. There is indeed only one way to have a total score of three, as each die must show one; but there are three ways to have a total score of four, as any one of the three colours can show a Two, with the other two dice showing One. So a score of four is actually three times as likely as a score of three. The same counting applies to outcomes such as (4, 4, 2), when two of the three dice show the same, with the third different. When all three are different— for example, when nine arises through (5, 3, 1)—there are six ways this can happen: any one of the three colours can show Five, combined with any one of the two remaining showing Three, leaving the last colour to show One.

Returning to the original problem, we now see that a total of ten is more likely than nine. The numbers of ways the totals can arise are:

$$(\text{for } 9): 6 + 6 + 3 + 3 + 6 + 1 = 25$$

$$(\text{for } 10): 6 + 3 + 6 + 6 + 3 + 3 = 27.$$

This exercise in counting has a distinguished history. Besides building telescopes, founding dynamics, and upsetting the church with his advocacy of the Earth's rotation of the Sun, Galileo found time to study gambling. He wrote a small treatise on the subject, prompted by a request from a group of Florentine noblemen to explain why this total of ten arose more often than nine. Table 5.4 shows how many ways all the totals for three dice can occur. If two totals (e.g. 4 and 17, or 10 and 11) add up to 21, their chances are exactly the same. To find individual probabilities, divide by 216. Totals of 9, 10, 11, or 12 arise very nearly half the time ($104/216 = 13/27$).

This is reminiscent of totals of two dice, where the middle value seven is most likely, and the probabilities fall away symmetrically as we move away from it. The same phenomenon occurs with totals for any number of dice: the most likely value is that in the middle of the range. With an even number of dice, this middle value is unique; with an odd number, such as the three dice here, the two middle values (10 and 11) are equally likely. By using ideas from number theory in pure mathematics, there is a way to write down a formula that will give the probability of any total using any

Table 5.4 Frequencies of the various totals when three dice are thrown. Frequencies for totals 11 to 18 obtained by symmetry.

Total	3	4	5	6	7	8	9	10
How often	1	3	6	10	15	21	25	27

number of dice; you can find it in Chapter 5 of Richard Epstein's *The Theory of Gambling and Statistical Logic.*

Four dice—Chevalier de Méré's problem

The systematic study of probability is often traced to seventeenth-century correspondence between the French mathematicians Blaise Pascal and Pierre de Fermat. Chevalier de Méré was a court figure and notorious gambler, who drew the attention of Pascal and Fermat to what seems at first sight to be a paradox. Are you more likely, or less likely, to have at least one Six when four dice are rolled together?

This question is answered in Appendix II: the chance of no Sixes is 625/1296, the chance of at least one Six is 671/1296, so the odds favour betting on a Six appearing. De Méré became interested in the corresponding game, in which *pairs* of dice are thrown, and the hope is that at least one of these pairs shows a *Double Six*: how many pairs of dice are needed so that the odds are in favour of a Double Six?

One possible line is to note that there are six outcomes with one die, and 36 outcomes with two dice; since throwing four dice is favourable in the first instance, and 36 is just six times six, then throwing 24 dice (since $24 = 4 \times 6$) should be favourable in the second case.

If only! Here is another instance where the arithmetic is quite correct, but the answer is wrong because the arithmetic is irrelevant to the problem at hand. The correct argument with four dice comes from working out the chance of *no* Six. Use the same method with 24 rolls of a pair of dice. On one roll, the chance of a Double Six is 1/36, so the chance of no Double Six is 35/36. Hence, on 24 rolls, the chance of *no* Double Six at all is $(35/36)^{24}$, which works out at just above one-half (about 0.5086). You are slightly more likely to have no Double Six than at least one. Making the same calculations with 25 pairs of dice, the chance of no Double Six is just under one-half (about 0.4945), so the odds favour a Double Six when you throw 25 dice—but not when you throw 24.

The consensus is that de Méré was a good enough mathematician to know that the original argument suggesting 24 pairs of dice are enough was erroneous. The probabilities with 24 pairs of dice and 25 pairs of dice are so close to one-half that it is not plausible to suppose (as is sometimes written) that de Méré discovered, via his gambling losses, that the bet is unfavourable with only 24 throws. This tale should reinforce the warnings that probability is a subject full of seductively plausible arguments, which can fall apart when you examine them. You need a consistent framework, and valid rules for combining probabilities, or you will reach false conclusions that could cost real money.

Bettering an opponent

This game will remind you of Penney-ante, in a different setting. There are four dice, with their six faces numbered as shown:

> Red Die: 4, 4, 4, 4, 4, 4
>
> Blue Die: 8, 8, 2, 2, 2, 2
>
> Green Die: 7, 7, 7, 1, 1, 1
>
> Yellow Die: 6, 6, 6, 6, 0, 0

Since the sum of the scores on all the faces of any single die is 24, the *average* score on each die is 4. If you throw any of them a large number of times, the total scores will be roughly equal. No die has an advantage if you are seeking as large a total as possible over 50 or so throws. (The Red die is plainly very boring; you know the result even before you throw it.)

You can choose any one of the dice, I then select one of those that are left, and we throw our dice *once* each. The winner is the one with the higher score.

Set up like this, I shall win two-thirds of all games. The reason is that, whichever die you choose, I can find one of the other three that wins the contest with probability 2/3. The justification is straightforward.

- If you choose Blue, I shall select Red: Red always scores four, and two-thirds of the time Blue scores only two.

- If you choose Green, I go for Blue: one-third of the time, Blue scores eight and must win. The other two-thirds of the time, Blue scores two, and half *this* time Green scores only one, so Blue wins. Adding these together gives a total chance of 2/3.

- You try Yellow, I take Green: half the time, Green scores seven and must win. The other half of the time, Green scores one, and one-third of *this* time Yellow scores zero—so Green wins. And once again, the chances total 2/3.

- You go for Red, my choice is Yellow: Yellow wins whenever Yellow scores six—which occurs 2/3 of the time.

Diagrammatically,

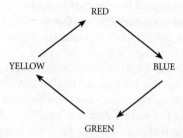

This shows the folly of arguments such as 'A is more likely than B, and B is more likely than C, so A is more likely than C'. Using such an argument for this game, you could make the blunder of musing: 'Red is better than Blue, and Blue is better than Green and Green is better than Yellow. So Red must be better than Yellow.' If you think along those lines, drop round for a game any day.

Fairground games

Try the following to raise money at the church fete. Chalk the numbers one to six in squares where gamblers can place their bets. You will roll three dice, and will pay out according to how often the chosen number appears. So if I place £1 on Four, and the dice show {6, 2, 3} I have lost my stake; if the dice had shown {4, 2, 2} I would get £2 (my original £1, + £1 for one Four); with an outcome of {6, 4, 4} I get a total of £3, and if all the dice showed Four, I get a total of £4.

At first sight, this may appear to be a money-loser rather than a money-winner. It seems that as there are three chances of winning, and six faces, then three times in six, or half the time, there will be a match and a loss of at least £1. But this is another example of loose thinking reaching the wrong answer. A valid analysis will note that the chance of *no* match in three dice is $(5/6)^3 = 125/216$, and not one-half as was suggested. The chance of one match turns out to be 75/216, two matches has probability 15/216, and all three dice match the selection with probability 1/216. The *average* return to a £1 stake is just over 92p.

Any church fete game that returned 92% of stakes would be dangerously generous! Although the odds are in the organizers' favour, there is too big a risk that random chance will lead to a loss. Participants in gambling games in church or similar fetes expect to lose money, so organizers should happily lever the game even more in their favour by simply paying out £2 to a £1 winning stake, irrespective of whether there are 1, 2, or 3 matches. The average return drops to a still-generous 84p. The bettor wins 91 times in 216, and loses 125 times.

A different game, using five poker dice, might be offered. Poker dice are ordinary dice, except that the faces are labelled Nine, Ten, Jack, Queen, King and Ace. Invite the punters to select two different faces that they hope will show when the five dice are thrown. If both faces named by a punter appear, s/he is paid at even money. The bet is lost if only one of the two named faces, or neither face, appears. This set-up is reminiscent of a popular bet at greyhound racing, where punters seek to name the first two dogs to finish, in either order. In our game, there would be no winners at all if all five dice showed the same outcome, but lots of winners when all five dice are different.

This game could only be considered as a money-raising prospect if the average return to punters is less than their stake. To find this average, fix attention on a bet that an Ace and a King will appear. Plainly, the probabilities will be the same for any other specific pair. The distinct possibilities are:

(1) neither Ace nor King appears

(2) just one of Ace and King appears

(3) both appear.

For (1), the chance that neither face appears on one die is $4/6 = 2/3$, so the chance that neither Ace nor King appears on five dice is $(2/3)^5$.

For (2), we will first calculate the probability that there is no Ace, but at least one King. The chances of no Ace is $(5/6)^5$. In those sequences with no Ace at all, the other five faces each show with equal probability, so the chance that no King at all appears is $(4/5)^5$. Collecting these together, the chance of no Ace, but at least one King, is:

$$(5/6)^5 \times (1 - (4/5)^5). \qquad (*)$$

Plainly, the chance of no King and at least one Ace is the same as this, so the total chance of (2) is just twice the value of $(*)$.

Having found (1) and (2), the easiest way to find (3) is to use the fact that all three probabilities must add up to 1. So the chance that (3) occurs reduces to:

$$1 + (2/3)^5 - 2 \times (5/6)^5 = 0.3279.$$

This is the chance that someone betting on a specific pair wins; it is just less than one in three. As you pay out even money, you can expect a good profit. If your punters are happy to try their luck with poker dice in this way, you might try to sell all 15 pairs of two different faces, such as (9, 10), (9, J), ... , (K, A) at £1 a time. The worst that can happen is that all five dice show different faces, and you pay out £20 in prizes, a loss of £5. But this happens less than one time in ten, and the rest of the time you are in some profit. Your average payout to a £15 stake is £9.84. Very occasionally—the one time in 1296 that all five dice show the same—you will pay out nothing at all!

Where do those estimates come from? For all five to be different, the second must differ from the first, the third from the first two, and so on. This leads to the expression $(5/6) \times (4/6) \times (3/6) \times (2/6) = 5/54 \approx 0.0926$ as the chance that all five dice show different faces, and you make the £5 loss. For the average return, we saw that you will pay out £2 to any punter with probability 0.3279. Thus the average payback on the 15 pairs is $£(15 \times 2 \times 0.3279) = £9.84$. This is less than 66p per £1 staked.

The same game could be played with any number of dice, but the more dice you use, the riskier. With six dice, you still expect to make a profit, as the average payback is 84p to £1 staked. Do not offer the game with seven or more dice, unless your aim is sabotage: these games favour the punter, and the church fete can expect to make a loss.

CHAPTER 6

....

Games with few choices

You and I require one ordinary coin each to play the following game. Each of us decides, in secret, whether to show Heads or Tails. We may reach our decision by any means we want: we could toss our coin, to let it make its own decision, or we may use whatever criterion we like to make a deliberate choice. When we have decided, we compare choices. If they are the same, you win two units from me; if they are different, I win two units from you. What these units are will be agreed before the game—they may be counters, pence, Euros or whatever. Is this game fair, or is it advantageous to one of us?

Such games are often best displayed in the form of a table. My possible choices are listed in the left-hand column, yours are shown in the top row. By changing the direct statement 'I lose two units' into the slightly convoluted 'I win −2 units' then all the entries in Table 6.1 show the outcome from my perspective.

There is no catch here. This game looks fair, and is fair. Provided each of us conceals our choice from the other, we are reduced to pure guessing. Half the time we guess right, and win two units, otherwise we are wrong and lose two units. The game is similar to the game of rock, scissors and paper, except that a draw never occurs.

Change the payouts slightly to those shown in Table 6.2. Is the game still fair? The answer is not clear cut. The rewards of three and one seem to be equivalent to the previous rewards of two and two, but you may hold some doubts. One possible line is for you to argue 'No, it is not fair. You will show Tails all the time, so I only win one unit, but you win two units.' I

Table 6.1 My payoffs in the basic coin-matching game, according to tactics used.

	You Heads	You Tails
Me Heads	−2	+2
Me Tails	+2	−2

Table 6.2 New payouts, same tactics as in Table 6.1.

	You Heads	You Tails
Me Heads	−3	2
Me Tails	2	−1

would counter that by pointing out that it would be silly of me to show Tails all the time, as you would respond by also showing Tails, and would win one unit every game.

Even if an argument is unsound, its conclusion may be absolutely correct. So it is here. The second game is *not* fair, because I have a way of playing that ensures that, whatever you do, I win 1/8 of a unit each game, on average. What I should do, and why it leads to this result, will have to wait a while. First, I will describe a similar 'game' in which the players do not exchange money or tokens, but each is seeking to maximize a reward.

The dating game

Darren has arranged to meet Zoe about eight o'clock outside the ZiPP Club. If both of them are early, or both are late, a happy evening is in prospect, to which Darren assigns a net worth of + 1, in suitable units that express his enjoyment. But if he is early, and she is late, he is likely to have an embarrassing encounter with his former girlfriend Ruth, which will ruin the evening. Overall, a net worth of −1 seems about right. Finally, if he is late, and Zoe is early, her former boyfriend Paul may nip in and take up with her again. From Darren's standpoint, this is even worse, and he gives this outcome a net worth of −2. He will make his decision on the basis of his 'payouts'.

	Zoe early	Zoe late
Darren early	1	−1
Darren late	−2	1

No money will change hands, but Darren's intention is to maximize his evening's pleasure. Zoe knows the score, and she will also consider whether to arrive early or late. She will have her own way to measure her prospective enjoyment, and could have a different opinion about the relative merits of the four possible situations. Her table of payouts may be quite different from Darren's, but that is irrelevant to him. He will play according to his table, as that expresses his view of the world. You may think Darren is being selfish, but his intention is to enjoy his evening out, so it is reasonable to look to his own interests. Zoe too should look at the four scenarios from her standpoint, and act as she thinks best. If she and Darren are incompatible, they will find out soon enough.

This dating game, and the coin matching game of Table 6.2, can be fully analysed in identical manners. But first, we look at a similar type of game that turns out to be easier to deal with. This new game will show why the other games need a different approach.

An easier game

Rather than describe games as a contest between me and you, I will invoke two new players: my role will be taken by Roy, yours by Colin (or Row and Column if you prefer). Roy's choices will be labelled as I and II, Colin's by A and B. It does not matter whether the two players have the same choices available since I, II, A, and B are just labels to distinguish each player's alternatives.

Roy's winnings are Colin's losses, and Roy's sole aim is to maximize his payoff. This conflict of interests is exactly the same as in the game of Table 6.2. In the dating game, it is a little different as Darren's gains are not Zoe's losses. Nevertheless, Darren simply seeks to maximize the payout according to his payoff table. Table 6.3 shows the payoffs to Roy in this new game. Roy fears that Colin may have spies who will discover which choice he will make, so wishes to take account of this too. Look along each row: if he uses I, he must win at least two units, whereas if he uses II, he can only guarantee to win one unit. Since using I leads to the larger guaranteed payout, he prefers I to II. Colin is equally fearful that Roy will discover his play, and considers his options for the two columns. Choosing A might lead to a loss of four, the worst that can happen if he chooses B is that he loses two. So Colin prefers B.

Table 6.3 Payoffs to Roy from Colin according to tactics used.

	Colin A	Colin B
Roy I	4	2
Roy II	3	1

They have each found a play, leading to a payout of two to Roy, that neither dare seek to improve on: if Roy switches to II, his payout may drop to one, if Colin changes to A, he might have to pay four. Their mutual interests are met by the choice (I, B), leading to Roy receiving two units.

This cosy arrangement does not apply to the game in Table 6.2 (or Table 6.1, for that matter). For Table 6.2, if I use Heads, I might 'win' only −3, but if I use Tails, I am guaranteed at least −1. This leads me to prefer Tails. Whichever you choose, you might have to pay two. Your potential loss is not the same as my potential gain, so we have no mutual interest that homes in on the same outcome. Similarly in the dating game. Darren sees that if he arrives early, he might end up with −1, and if late, it is possible he gets −2. The better of these is −1. Whatever Zoe selects, a payout of 1 is possible—and this is different. These games are more complicated than that in Table 6.3.

The steps to reach the choice of (I, B) in the game between Roy and Colin were:

(1) Find the minimum number in each row, and note the *largest* of these minima. Call it (ugly word) the *maximin*.

(2) Find the maximum in each column, and note the *smallest* of these maxima. Yes, this is the *minimax*.

As they turned out to be the same, our task was complete. Roy gained two. In most games of this type that are set up, the maximin and minimax will be different, but it is an excellent habit to check this point before you go any further. You might save an awful lot of time and trouble. This analysis even extends to games where the players may have several choices, as the next example shows.

A fisherman's tale

Roy and Colin are fond of fishing, and will arrange to meet at one of 12 lakes for a relaxing Whit Monday. However, Roy wants to catch pike, while Colin is a trout man—and more pike mean fewer trout. Their interests are in conflict. The lakes are situated at road junctions, and Roy will select the east–west road, while Colin chooses the north–south one. The number of pike per hectare of lake are shown in Table 6.4.

Roy looks along each row: the respective minima are 1, 4, and 2, so if he chooses route II, he guarantees a pike density of at least 4. Colin peers

Table 6.4 Numbers of pike/hectare in
each of 12 lakes, identified by roads.

	A	B	C	D
I	3	7	2	1
II	5	5	4	6
III	4	2	3	8

down each column, noting their maxima as 5, 7, 4, and 8. His best protection comes from road C, as the worst outcome there is 4 pike/hectare. They have a mutual interest in selecting (II, C).

When they have come to this conclusion, they no longer need to keep their strategy secret. Roy can announce he will choose II, safe in the knowledge that Colin can do no better than select C; and Colin can tell Roy that C is his choice, without Roy being able to gain any greater advantage. Their strategies are proof against spies.

If the numbers in Table 6.4 did not refer to fish, but were the heights of the lakes above sea level, in hundreds of metres, we have identified a lake at (II, C) that stands at a saddle point in the terrain. Everything to the north or south is lower, everything to the east or west is higher. The procedure of finding the minimax and maximin, and checking whether they are the same is akin to a surveyor noting saddle points among the mountains and valleys (of course, a surveyor is not restricted to just looking north–south and east–west). It is summarized by the phrase 'Look for a saddle point'.

This chapter is about games with few choices, so that drawing up a table of all alternative outcomes is feasible. Our fisherman's tale shows that the notion of a saddle point can instantly lead to a solution of a game where each player may have many choices. Even in games that have rather fewer choices, checking for a saddle point is the first step to take.

Back to Table 6.2

That game has no saddle, and I claimed that I had a way of winning 1/8 per game, on average. Let Roy and Colin take over from me and you. In these games without saddle points, a player should consider using each of his choices every time, and switch randomly between them. The key step is to decide the relative frequency for each choice. In this game, it turns out that Roy should have a bias towards T. Specifically, he selects H with probability 3/8 and T with probability 5/8, independently each game. This notion is called a *mixed strategy*.

To see why that is a good idea, look what will happen when Colin chooses H. This means we focus on just the first column of Table 6.2. In these games Roy 'wins' −3 with probability 3/8, and wins 2 with probability 5/8, so his average winnings are

$$(-3) \times \frac{3}{8} + (2) \times \frac{5}{8} = \frac{1}{8}.$$

When Colin selects T, we use the second column of Table 6.2. Using the same argument, Roy wins, on average,

$$(2) \times \frac{3}{8} + (-1) \times \frac{5}{8} = \frac{1}{8},$$

again. Whether Colin chooses H or T, Roy's average winning are 1/8. And just as in the fisherman's story, Roy could tattoo his strategy on his forehead—Colin could do nothing to prevent the advantage. Over eight games, Roy's average winnings are one unit, over 80 games he gains ten units on average. He cannot guarantee to win, but the game favours him.

We need a practical way for Roy to achieve his object of showing H or T at random, with the correct frequencies. There are two things he is likely to carry that will be useful. First, the stopwatch facility on a digital watch usually gives timings to one-hundredth of a second. Start the stopwatch, and when it is time to make the choice of H or T, press the stop button and look at the digit in the *second* decimal place. That digit must be one of $\{0, 1, 2, \ldots, 9\}$, and I am prepared to believe they are all equally likely. If this second digit is 0 or 9, restart the stopwatch, and stop it again a few seconds later. Sooner or later, the final decimal digit will fall among $\{1, 2, \ldots, 8\}$. When it does, show H if the digit is 1, 2, or 3, and T otherwise. This gives H and T with the right frequencies, independently each time. Let the stopwatch keep running between games.

A second method is to use three coins to make the decision. With three fair coins, we know there are eight equally likely outcomes, and that just three of these contain exactly one H. So an alternate method is for Roy to retreat to a corner, toss three fair coins, and then show H in the game when exactly one of these three supplementary coins gives H. These external agencies, coins, or stopwatches are far better than humans at choosing independently and with the right frequencies!

Games players who wish to succeed must have their own randomizing devices. A stopwatch that reports values to one-hundredth of a second can be adapted to give any required frequency with good enough accuracy. Otherwise, clever work with coins, dice, or a pack of playing cards can ensure truly random choices, at any desired frequencies.

The fact that Roy's average winnings are only 1/8 may be beneficial to him in the long run, as Colin may take a very long time to realize that he is at a disadvantage. In Penney-ante (Chapter 4), the second player wins at least twice as often as the first, so the first player may drop out early. Here, if Colin chooses H and T at random and equally often, they each win about the same number of games, but Roy pays Colin one unit rather more often than he pays three. His advantage is much less noticeable.

Suppose Roy uses this strategy over a series of games. How likely is it that he will be ahead? The answer depends both on Colin's tactics, and on the number of games. For example, suppose Colin shows H every time. This simplifies matters, as there are only two possible outcomes for each play: either Roy wins two, or he loses three. Figure 6.1 shows all the possibilities after one and two plays.

What Roy plays is shown next to the arrow. After each play, beside the arrow head, are the probabilities and his corresponding total winnings. There is a nice curiosity after two plays: although Roy's average total winnings are 1/4 (made up from two lots of 1/8), Colin is the more likely to be ahead. Roy is ahead just 25/64 of the time, Colin leads the remaining 39/64 of the time.

If Colin shows H and T at random, matters get more complicated as

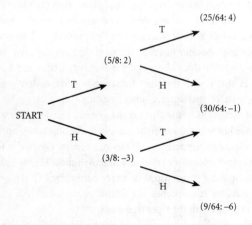

Fig. 6.1

there are now *three* outcomes possible. In half of all games Roy wins two units, in 5/16 of them he loses one unit, and in 3/16 of them he loses three units. Without giving the details, it transpires that after two games not only is Roy 1/4 unit up on average, he is now more likely than Colin to be ahead.

What about much larger numbers of games, when Colin desperately shows H or T at random, equally often? How likely is it that Roy will be ahead after 50 or 100 games? And how many games are needed so that Roy's chance of being ahead is over 90%? The calculations that would lead to exact answers are not difficult, but they are very, very tedious, and a much better approach is to use simulation on a computer.

Table 6.5 gives estimates of the probability that Roy is ahead, or level, after a given number of games. Each estimate is based on simulating a million plays of that series, which gives me great confidence in the accuracy of the figures. After 16, 32, 64, ... games, Roy is ahead by 2, 4, 8, ... units, on average. But he could well be losing even after a long series—his chance of being behind after 500 games is about 8%.

Table 6.5 Estimates of the probabilities that Roy is ahead, or level, after different numbers of plays of the game in Table 6.2.

Games	16	32	64	128	256	512	1024
% ahead	58	62.5	69	76.6	84.5	92	97
% level	5	3.3	2	1.2	0.7	0.3	0.1

A fair game?

The notion of fairness is not relevant in the dating game where there is no exchange of money or tokens. But where there is direct conflict between the two players, it is useful to see how we could make games fair, by ensuring that the average winnings to either Roy or Colin are zero. That was already the case for the matching coins game of Table 6.1, but not for that of Table 6.2. That game can be made fair if the favoured player pays a suitable amount before each contest to the other one.

For the Table 6.2 game, Roy can ensure an average win of 1/8 every game. Moreover, that is the best he can do—there is no other play that guarantees a higher average payout against best play by Colin. This game will become fair if Roy makes a side payment of 1/8 to Colin for every game played.

Games with saddle points are even easier. The value at the saddle is the appropriate payment. For the fisherman's saga, where the 'rewards' to Roy are not expressed as monetary amounts, we have to reformulate the game in some suitable currency. The higher the pike density, the more Roy desires that particular lake, and the less Colin does. They should negotiate a value for fishing at each of the 12 lakes, expressed as a payment from Roy to Colin. The payments at the lakes such as (I, D) or (III, B) will be negative, as Colin is anxious to fish there, while Roy will pay most for the lake at (III, D) with eight pike /hectare. Whatever these new numbers are, the saddle point will remain at (II, C), since all that matters is the relative *order* of the numbers in the table. The game is now fair when Roy pays the corresponding sum.

Finding the best strategies

For any game in which each player has two choices, and there is no saddle point, there is a simple recipe to find the best strategies. It is easiest to describe with a numerical example, so return to Table 6.2, slightly modified. Table 6.2A is the same as Table 6.2, but with ten added to each payoff. This changes the game considerably—Roy's reward is ten more, whatever the choices—but the main effect is to get rid of the distracting negative numbers. We find Roy's best strategy first.

Table 6.2A As Table 6.2, but with all payoffs increased by ten units.

	Colin Heads	Colin Tails
Roy Heads	7	12
Roy Tails	12	9

Step 1: Subtract the numbers in the *last* column from the corresponding ones in the *first* column. This leads to:

$$\begin{matrix} -5 \\ +3 \end{matrix}$$

Step 2: Ignore the minus sign, and swop the entries round. This gives

$$\begin{matrix} 3 \\ 5 \end{matrix}$$

Step 3: Choose the two strategies in proportion to these numbers. Here, since $3 + 5 = 8$, Roy chooses H and T with probabilities 3/8 and 5/8 respectively.

Colin also has a best strategy, guaranteed to protect him even if Roy's spies discover it. This is found by turning the recipe for Roy on to its side.

Step 1: In the same way, subtract the numbers in the *bottom* row from those in the *top* row. This leads to: $\boxed{-5 \quad 3}$

Step 2: (As Roy) Ignore the minus sign, and swop them round. Here $\boxed{3 \quad 5}$

Step 3: (As Roy) Choose the two strategies in proportion to these numbers. In this game, Colin should also select H or T with respective probabilities 3/8 and 5/8.

(It will not always be the case that both players have the same best strategy.)

Remember that Table 6.2A arose by adding ten to every entry in Table 6.2. Aha! If you add ten to every entry, then *subtracting* the corresponding entries in the rows or columns leads to the same answers in the boxes! So even if we had started with Table 6.2 itself, we would have reached the same conclusion. Check it if you like—here is Table 6.2 repeated for convenience.

Table 6.2 (As before).

	You Heads	You Tails
Me Heads	-3	2
Me Tails	2	-1

Subtracting the numbers in the last column from the corresponding ones in the first column gives –5 and then +3, as before. So Steps 2 and 3 lead to the same mixed strategy of choosing H or T with respective probabilities 3/8 and 5/8.

To confirm that you are happy with this recipe, work out the best strategies when the rewards to Roy in the matching pennies game are as in Table 6.6. You should discover that Roy selects H one time in four, while Colin chooses both H and T equally often.

Now that we have a way to find the best strategies, calculating Roy's average winnings comes in one more step. Find the average of the entries in the first column, averaged according to the weights in Roy's best strategy. For Table 6.2, as we saw earlier, this average is $(-3) \times (3/8) + (2) \times (5/8) = 1/8$. The same answer should arise when you average the second column in the same way—a useful check. The arithmetic for the game in Table 6.6 is similarly $(-3) \times (1/4) + (1) \times (3/4) = 0$: this game is fair.

The average amount Roy wins is termed the *value* of the game. An unfair

Table 6.6 Rewards to Roy from Colin according to tactics used in matching coins.

	Heads	Tails
Heads	−3	3
Tails	1	−1

game is made fair by Roy paying Colin this value. Recall that Table 6.2A arose by adding ten to all Roy's payoffs in Table 6.2. Colin would require some compensation for this change, and if Roy made a side payment of ten units every game, Colin could happily agree the change. The value of Table 6.2A, as it stands, works out to $10 + 1/8$, so now all is reconciled: the 6.2A game is fair when Roy makes a total payment of $10 + 1/8$, every game.

Adding the same amount—ten in this case—to every payoff will change the value of a game, but always leaves the best strategies for both players unaltered. The same applies if you multiply all payoffs by the same (positive) number. So you can often simplify your arithmetic to locate the best strategy by the two following steps:

(1) Eliminate fractions by multiplying all payoffs by a suitable positive number.

(2) Eliminate negative numbers by adding a suitable amount to everything.

When you have found the best strategies, go back to the original payoffs to find the value of the game.

Other games

Let's put our knowledge of how to find the best strategies, and whether the game is fair, to the test by setting up and solving some games. The essential element is that the two players are in conflict, with Rows wishing to maximize.

The dating game

As a reminder, Darren stares at this conundrum.

	Zoe early	Zoe late
Darren early	1	−1
Darren late	−2	1

We saw that this has no saddle point, so Darren will use a mixed strategy. To find the respective odds, follow the recipe. Subtracting the second column from the first gives the values $(2, -3)$. Ignoring the minus sign and swapping them round leads to proportions of $(3, 2)$, so Darren should arrive early with probability 3/5, and late with probability 2/5. The average pleasure Darren will obtain is the value of the game, i.e. $1 \times (3/5) + (-2) \times (2/5) = -1/5$. Negative pleasure is unattractive. Perhaps Darren should cancel the date, or reconsider his payoffs!

Perhaps Zoe's table of payouts has the form

	Darren early	Darren late
Zoe early	1	3
Zoe late	2	1

(She's going to enjoy herself anyway; she'll get extra pleasure if Darren has to bump into Ruth, and Paul was much more fun!) Column subtraction gives $(-2, 1)$, which transforms to $(1, 2)$, so she should arrive early with probability 1/3, or late with probability 2/3. The value—the average amount of pleasure she will derive—is then 5/3.

On the other hand, suppose she has a very different attitude towards Paul, and change the entry '3' to become '-1'. Her new table of payouts has a saddle point, and she should automatically turn up late. That makes sense.

The guilty husband

Roger has forgotten whether it is Christine's birthday. Should he buy her a bunch of flowers to present as he arrives home? This theory should help.

Suppose it is not her birthday. If he brings no gift, that is a neutral position, with a baseline payoff of zero; but if he surprises her with the flowers, she will be pleased in any case—a payoff of, say, $+5$. But suppose it *is* her birthday. Bringing the gift establishes he has remembered, so a higher payoff is appropriate—maybe $+8$. Worst is no gift, leading to calamity, expressed in this scale as -20. Still short of the florists, he forms his payoff table. He is not competing against Christine, his enemy is perverse Nature that either did or did not select today as her birthday.

	Not birthday	Birthday
No gift	0	-20
Flowers	5	8

As ever, look for a saddle point. The smallest row values are −20 and +5, the larger of these is 5; the largest column values are 5 and 8, the smaller of these is also 5. There *is* a saddle point, so Roger should *buy the flowers anyway, whether or not it is his wife's birthday.* (Wives should leave their copy of this book open at this page for their husbands to read.)

Military decisions

Few military decisions are based on the notion that you and your enemy each have just two strategies to choose from. But it could be realistic to think in terms of both sides choosing between five or six strategies. For example, a commander may have to defend two territories using four divisions, so his five choices are to deploy them as (4, 0), (3, 1), (2, 2), (1, 3), or (0, 4) at the two sites. To work out a best strategy for a payoff table when each contestant has three or more choices is much more complicated than when one side has only two alternatives, and I will not include such examples. But the final chapter of John Williams' *The Compleat Strategyst* describes how to proceed in those cases. My debt to that entertaining book for some of the examples in this chapter is considerable.

Two midget submarines seek to get close enough to the enemy's aircraft carrier to launch a successful torpedo attack. One submarine carries the torpedo, the other is packed with radar-jamming devices and diversionary dummies. They must pass through a guarded channel, too narrow to be side by side, so one submarine must be ahead of the other. The enemy will have an opportunity to attack one submarine only; the chance the first submarine in the channel will survive an attack is 80% , but that for the second is only 60%, as the enemy have more warning. If the submarine carrying the torpedo gets through the channel, it is certain to succeed in its mission. Which submarine should carry the torpedo, and how likely are we to destroy the aircraft carrier?

For this problem, an appropriate payoff is the *percentage probability the carrier is destroyed*. Strategies I and II are to place the torpedo in the first or second submarine respectively, the enemy's choices A and B are to attack the first or the second. From our standpoint

	Attack on first	Attack on second
Torpedo in first	80	100
Torpedo in second	100	60

which has no saddle point. Using our algorithm (the mathematical word for a recipe that leads to the answer) on the two columns, the first difference is $(-20, 40)$ which translates to proportions $(40, 20)$. We should put the torpedo in the first submarine with probability 2/3, and in the second with probability 1/3.

The value of this game is $80 \times (2/3) + 100 \times (1/3) = 260/3$, so the chance of success is nearly 87%. On the same data, the enemy should attack the first submarine with probability 2/3, but that is for them to discover.

A naïve naval officer, untutored in these matters, might conclude that since 80% exceeds 60%, he should automatically put the torpedo in the submarine that will go first through the channel. If he did so, the chance of success is simply 80%. Naval officers who, through their misplaced ignorance reduce the chance of a mission's success from 87% to 80% should be cashiered and moved to less important duties.

The Spoilers (1)

John Stonehouse, a fallen politician fleeing justice, is said to have discovered how to create a new identity for himself from Frederick Forsyth's *The Day of the Jackal*. Desmond Bagley's pot-boilers are read less than Forsyth's thrillers, but they too can be useful in their supplementary information. In *The Spoilers* explicit use of the solution of games of two strategies is used near the end, as the heroes seek to decide best how to escape from the villains. The book also has another episode in which superior mathematical skills enable one character to lure another into an unfavourable game (see Chapter 7). Imaginative maths teachers could do worse than prescribe *The Spoilers* as homework early in Year 11. (It also brings in the coin-tossing game of Table 6.2.)

In honour of Bagley's enterprise, I offer a similar scene for another fictional character:

> Bond did not even know the day of the week. Disoriented from his incarceration, he struggled to the jetty. His mind cleared as he saw the speedboat still there. Should he go north or south to escape Blofeld's certain pursuit in the faster craft? Safety was closer to the south, and he might not have the fuel for the northern route. Blofeld would know all this too, but Bond's mind was now clear: he jotted down some figures on a scratchpad, and stopped. Damn! No watch. His steel-blue eye caught the transmitter, its

battery low. Desperately he pressed the button, and got the response immediately. 'Q here; where the hell are you 007?' 'That's irrelevant Q: just tell me the day of the week.' 'Monday; England are 160 for 7, hanging on for a draw.' Bond heeded the response: Monday. He turned north.

The explanation rests, as you will have realized, on Bond's payoff table. Examining each option in turn led to these assessments of his percentage probability of escape.

	Blofeld south	Blofeld north
Bond south	70	100
Bond north	80	40

The usual algorithm leads to Bond's best strategy of (4/7, 3/7) to south and north respectively. Having no randomizing device handy, he mentally assigned Sunday, Monday and Tuesday to north, and called Q. His chance of escape is just over 74%. (Blofeld should go south with probability 6/7.)

The prisoners' dilemma

Roy and Colin have robbed a bank together. The police know this, but cannot prove it, and are holding them in separate cells while seeking more evidence. The two could clearly be convicted on lesser crimes, but the police want to nail them for the bank job. Their intention is to persuade either prisoner to give evidence to convict the other, and to gain such evidence they would be prepared to drop the lesser charges.

Roy has the following dilemma. 'If we both keep quiet, we will both get two years for the lesser crimes. If I rat on Colin, but he keeps mum, I go free and he gets six years for the robbery—and vice versa. If we incriminate each other, we can both expect five years—six years less a year for helping the police. All these "rewards" are negative. Colin is my friend—I think— so by "co-operation" I mean keeping quiet, and by "defection" I mean giving evidence. Here is my payoff table; I'll add six to everything to get rid of the negative numbers and make life easier.'

	Colin co-operates	Colin defects
Roy co-operates	4	0
Roy defects	6	1

Look for a saddle point: yes, there is one when both defect! So the 'solution' is that both defect—and end up with five years in jail!

The moral of the tale is to love and trust thy fellow man: here Roy and Colin are *not* in direct conflict with each other. But if they act selfishly in their own best interests, thinking the worst of the other, they are both worse off than if they had shown honour among thieves and kept quiet. If both kept quiet, the outcome would have been the much preferable two-year sentence.

Tosca

Anatol Rapoport modelled the plot of Puccini's opera, *Tosca*, as an exercise in game theory. The contestants are the beautiful Tosca, and the chief of police, Scarpia. Scarpia has condemned Tosca's lover, Cavaradossi, to death, but offers to reprieve him in exchange for Tosca's favours. Each of Tosca and Scarpia has the chance to doublecross the other. Scarpia can opt to use blank or real cartridges in Cavaradossi's firing squad, Tosca has a concealed knife that she might use when Scarpia embraces her.

Tosca's ideal is to have her lover reprieved and Scarpia dead. But Scarpia's first preference is to have Cavaradossi out of the way, and Tosca all to himself. Tosca's thoughts run: 'If Scarpia really does use blank cartridges, I'll be much better off if I kill him, then I can be with Cavaradossi. But if Scarpia uses real ones, I'll at least get some revenge if I kill him. In some suitable units, my payoff table is:

	Scarpia plays fair	He double crosses
She plays fair	5	−10
She kills Scarpia	10	−5

Whichever choice he makes, I prefer him dead.'

Scarpia too assesses his options. 'If she trusts me, I want to have her lover out of the way. But if she kills me, it will be some consolation that she no longer has him to turn to. My table of payoffs is:

	She plays fair	She stabs me
I keep my word	5	−10
I kill Cavaradossi	10	−5

Whatever she does, I should execute her lover.'

This is not a joyous tale. Both Tosca and Scarpia argue that whatever the other does, to doublecross them is preferable to playing fair. So both achieve a disastrous result: Scarpia is dead, Tosca's lover is dead. The payoff tables have different numbers from the table for the prisoners' dilemma, but the essential reasoning is the same. In each case, lack of trust dominates. The worst of all outcomes is when you play fair, but the other does not. To prevent this, you get your retaliation in first.

Two choices for Roy, many for Colin

Such games can be solved almost as easily, and using the same methods as we have seen. The first step is the standard one of looking for a saddle point, and if you find one, there sits the solution. Otherwise, you might be able to simplify matters by eliminating some of Colin's choices that are plainly worse than others. Remember that Roy wants big payoffs, so Colin expects to concentrate on those selections where payoffs to Roy are *small*. For example, suppose the payoffs are:

	A	B	C	D
I	2	3	6	7
II	5	6	3	4

There is no saddle point, but compare C and D from Colin's position. Whether Roy chooses I or II, Colin is better off with C, and so strategy D can be struck off—a rational Colin would never use it. Similarly, compare A and B; Colin prefers A whatever Roy selects, so B can also be eliminated. This leaves:

	A	C
I	2	6
II	5	3

which we know how to solve. Roy should use I or II with respective frequencies 1/3 and 2/3, while Colin uses A and C equally often.

This idea is described by the word *dominance*. For Colin, C dominates D and A dominates B, so it is never sensible for him to use B or D. So if you find no saddle point, look to throw out as many dominated strategies as possible. If you end up with two strategies for each player, you know what to do.

This is the place for a small digression. Where Roy and Colin have just *two* choices each, the only way there will be a saddle point is if at least one of the players—and possibly both—has a dominated strategy. So an alternative way of getting to the same solution is first to eliminate a dominated strategy, and then the solution is obvious. This was the case with the guilty husband: buying flowers dominates not buying them—he is always better off. Also in the prisoners' dilemma 'defect' dominates 'co-operate' for both players. However, when there are more than two choices around, saddle points can arise without there being dominated strategies (see the fishing story), so we have to consider both ideas. Back to the main theme.

Counter signals

Is the following a fair game? Roy has two counters, and shows either one counter or both of them. Colin has three counters, and must show either one, two, or all three. They make their choices in secret, and then reveal them together. Whatever the total number of counters, that amount changes hands: Colin pays Roy if the total is even, Roy pays Colin when the total is odd.

Setting this up via a table of payoffs to Roy, we find:

	Colin shows 1	Colin shows 2	Colin shows 3
Roy shows 1	2	−3	4
Roy shows 2	−3	4	−5

There is no saddle point, and no obvious dominated strategy. We are in new territory for seeking a solution. The saving grace if we reach this stage is that there will *always be a best strategy for Colin that uses only two of his choices*. Given this key fact, I will describe a logical process by which the best strategies must emerge, and then show a simple diagrammatic method of getting there, with even less calculation.

This logical approach is to look at all the possible games from this set-up where Colin deliberately restricts himself to just two choices. There are three such games (Colin can use his choices $(1, 2)$, $(1, 3)$, or $(2, 3)$), so we now use our accumulated wisdom to find Roy's best strategy for each of these three games separately. This gives (up to) three candidate strategies for Roy. You are not impressed? I think you ought to be. Recall that Roy will use a mixed strategy, which is fully specified when we state the frequency he shows one counter. At the outset, there are *infinitely many* possibilities—any proportion between zero and one. Coming down from infinity to three is pretty good progress.

Now pick out one of the three mixed strategies for Roy. Find the average payoff he would get against each one of Colin's three possible choices, and write down the *smallest* of these. That is the most Roy can guarantee to win, on average, if Colin's spies are lurking about. Repeat this for all three of Roy's candidate strategies, and then find the *largest* of these three guaranteed winnings. That is the value of the game to Roy, and he should use the strategy that leads to it. Colin can go through the same calculations, and that will direct him to those two of his three choices that he should use. Finding the frequencies for these two choices is then standard.

This process is less fearsome than it may seem. The table below shows the outcome of these calculations for each of the three 2 × 2 games. To specify Roy's strategy, we simply write down in the second column the frequency with which he shows just one counter. The last three columns show Roy's consequent average winnings against each one of Colin's three choices.

Colin's choice	Prob. Roy chooses 1	Colin 1	Colin 2	Colin 3
(1, 2)	7/12	−1/12	−1/12	1/4
(1, 3)	1/2	−1/2	1/2	−1/2
(2, 3)	9/16	−3/16	1/16	1/16

Look along each row. If Roy shows 1 counter with probability 7/12, he will win at least −1/12 whatever Colin does; if he shows 1 with probability 1/2, he can only guarantee to win −1/2; and if his probability of showing 1 is 9/16, this ensures he wins at least −3/16. All Roy's guaranteed winnings are negative, so the game favours Colin. But the *largest* of these winnings is −1/12, and that is the value of the game. It becomes fair when Colin pays Roy a fee of 1/12 for each time they play. Roy's best strategy is to show one counter with probability 7/12, and two with probability 5/12. To counter this, Colin shows either one counter, or two counters: *he must never show three counters*. If he did so against Roy's best strategy, the game is already worth 1/4 to Roy, before the additional fee Colin pays to make it fair!

If Roy does use his best strategy, and Colin restricts his choices to one counter or two counters, the probability he chooses one, or chooses two, is irrelevant; his average gain is 1/12 each time. But if Roy deviated from his best strategy, Colin would wish to take advantage. We can find *his* best play

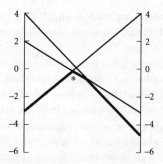

Fig. 6.2

by looking at the 2 × 2 game where he uses only (1, 2). By standard methods, he shows one counter with probability 7/12, or two counters with probability 5/12.

I promised a diagrammatic method that would be easier to implement. The diagram is shown in Figure 6.2. The left line corresponds to Roy choosing one counter, the right line to him choosing two counters.

The three diagonal lines correspond to Colin's three choices. For any one of these lines, the leftmost point gives Roy's winnings when he chooses '1' all the time, the rightmost point is when he chooses '2' all the time. Going along the diagonal line from left to right shows his average winnings as he increases the probability of using '2'. Work from left to right, highlighting the *lowest* of the three diagonal lines, as shown. Then find the *highest* point on the path you have traced. That will inevitably be where two diagonal lines meet, shown by *. This picks out the two strategies Colin should use if he plays optimally, and so identifies which one of the three 2 × 2 games we should solve.

Of course, if you use graph paper and can read off values on the scales, your arithmetic is already completed! Find the exact position of the highest point. Its value on the vertical scales is the value of the game, and how far it is along the path from the left scale to the right scale gives the probability that Roy shows '1'. In this diagram, the highest point is at −1/12 on the vertical scale, and it is nearer the left scale than the right scale in the ratios 7:5, so show one counter or two counters with probabilities 7/12 or 5/12.

One bonus from this diagrammatic approach is that dominated strategies are automatically taken care of. If Colin did have a dominated strategy that he could have eliminated, this will show up as a diagonal line that never gets highlighted as we move from left to right! But it is best to eliminate obvious dominated strategies anyway, to simplify as much as possible.

One way to assure yourself that you can implement these techniques is to extend this game, by allowing Colin to show up to four counters, instead of three. The table of payoffs will have one more column, with entries −5 and 6, and now Colin has three additional mixed strategies to use, i.e. (1, 4), (2, 4), and (3, 4). This might give three more candidate strategies for Roy, as he considers his best response against each of these new possibilities. In fact— the details are left to you—Roy's best response to (1, 4) is the same as his best response to (2, 3), which we have already found. This gives just two new candidate strategies, making a total of five.

To reach his decision, Roy has to see how Colin might fare against all these five strategies. This includes going back to the three we have already considered, as Colin has a new option—to show four counters—that might be better for him. In fact, if Roy did use his previous best strategy, showing one counter with probability 7/12, *Colin* will be at an advantage when he shows four counters, gaining 5/12 on average. When Roy completes all his calculations, he (and you) should find that the best he can do is average winnings of −3/16. It should be no surprise that this is worse than before, as Colin has an extra option. Roy's best strategy is to show one counter with probability 9/16. To make this new game fair, Colin's entrance fee is 3/16.

What should Colin do in this extended game? He should show either one counter, or four counters: *he should never show two or three counters.* Either of those choices favours Roy. By the standard algorithm, Colin should select one counter with probability 11/16, or four counters with probability 5/16.

The diagrammatic method of finding the best strategies in this extended game is very easy, if you have already used it when Colin had three choices. All that is needed is to draw in one further line, joining −5 on the left axis to +6 on the right axis. This new line cuts the diagonal line from +2 to −3 at a point marginally to the right of the old *, and slightly lower; this shows up as a lower value to Roy (−3/16 is less than −1/12), which he achieves by showing one counter a little less often than before (now 9/16 of the time, previously 7/12).

The necessity of a stopwatch or other suitable randomizing device should be clear by now. Without it, could *you* consistently make those fine decisions that ensure you show one or two counters with the right average frequencies, independently each time? I know I could not. Not using a randomizing device is giving an advantage to an opponent who does so—action anathema to all serious games players.

Games animals play

Everyone knows the central place of mathematical models in physics. Their use in biology is less well appreciated. But the operation of Mendel's laws is best seen mathematically, and great advances in biological understanding came through the models developed by R. A. Fisher, J. B. S. Haldane, and Sewell Wright. More recently, game theory has been used to understand animal behaviour.

One prototype is a model of contests between two animals who are disputing ownership of food, territory or mating rights—anything that will aid their success in passing on their genes to the next generation. The simplest idea is that just two types of behaviour can be used, Hawk or Dove. A Hawk is always willing to start a fight, and will fight back if attacked; it will retreat only when too severely injured to continue. A Dove will posture, but not begin a real fight; if attacked, it retreats before it suffers any damage. Two contesting Doves will huff and puff for some time, wasting time and energy, before one tires and concedes the prize.

The prize is measured in what biologists would call 'added fitness', a way of describing the extra average number of offspring they obtain from victory. Losers have offspring too; the prize just gives extra offspring, who can be expected to follow their parents' successful tactics. So if Hawk does better than Dove in these contests, we can expect the frequency of Hawks to increase, and vice versa.

To draw up a payoff table, we need some scale and reference point to describe the values of the prize being contested, the loss of fitness through injury in an escalated fight, and the loss of fitness through wasting time on display. Take the value of the prize as ten units, and assume that in Dove–Dove or Hawk–Hawk contests, either player is equally likely to win, so either gains five units, on average, from the contest. In a Dove–Hawk contest, the Hawk wins all ten units, the Dove wins nothing, but suffers no damage either. In Dove–Dove contests, deduct two units from each player to account for the fruitless time they seek to stare the other out, but assume Hawk–Hawk contests are so fierce that the average loss to both exceeds their average gain of five—make the net value –2. The payoffs to the player using the tactic in the left column are

	Hawk	Dove
Hawk	–2	10
Dove	0	3

This contest is quite different from all the contests we have looked at so far. We are seeking to discover which of the two available *tactics* does better, not how Rows should compete against Columns. The new approach is to assume that the population contains some proportion, p, of players who use Hawk, so that a proportion $1 - p$ will use Dove, and see how well each tactic does in such a population. If one tactic does better than the other, the use of the favoured tactic can be expected to increase.

When a Dove decides to enter a contest, the chance it picks a contest with a Hawk is p, with another Dove it is $1 - p$. So the average outcome to the Dove is

$$(0) \times p + (3) \times (1 - p) = 3 - 3p.$$

Similarly, the average payoff to a Hawk is

$$(-2) \times p + (10) \times (1 - p) = 10 - 12p.$$

This means the two tactics do equally well, on average, when $3 - 3p = 10 - 12p$, which is when $p = 7/9$. When the proportion of Hawks exceeds $7/9$, the average payoff to Hawks is less (they get involved in too many damaging fights), so Hawk numbers reduce. But when the proportion of Hawks is less than $7/9$, Hawks are favoured and their numbers tend to increase.

A population can collectively play Hawk $7/9$ of the time in many ways. At one extreme, $7/9$ of the population play Hawk all the time, and $2/9$ play Dove. At the other extreme, every individual uses the mixed strategy $(7/9, 2/9)$ all the time. In either case, an interloper has probability $7/9$ of facing a Hawk, and $2/9$ of facing a Dove, and that is all the information we used to compute this neutral position. Suppose a population does collectively use this strategy in some manner, but that a small mutant population playing Hawk at a different frequency arises. Can this mutant survive?

The mutants use Hawk and Dove in some proportions $(x, 1 - x)$, where x can take any value except $7/9$. When they are present in only small numbers, virtually all their battles are against the indigenous population. But since each of Hawk and Dove do equally well, on average, these mutants also gain the same average benefit from the games they play. They are at no advantage, or disadvantage, so sheer random chance determines whether their numbers increase or decrease. Sometimes random chance eliminates them anyway.

Should the mutants increase their numbers, more of their contests are against other mutants. Their fate will depend on a comparison of how well a mutant fares against another mutant with how well the indigenous population fare against mutants. And the arithmetic shows that the originals do *better*; as soon as the mutants begin to increase in numbers, they become at a disadvantage, and so can make no headway. They are inevitably eliminated, whatever the proportion who play Hawk.

John Maynard Smith coined the term *evolutionarily stable strategy*, ESS for short, to denote any strategy that will resist all invasions. Much more can be read about the use of game theory to comprehend animal behaviour in his *Evolution and the Theory of Games*. In this example, if Hawk and Dove are the only tactics available, then using them in proportions 7:2 is an ESS for this table of payoffs.

Change the numbers in the payoff table, and a different population composition may emerge. Make the item being contested less valuable, but suppose the Hawks fight more fiercely and inflict more damage on each other. The new numbers may be:

	Hawk	Dove
Hawk	-4	6
Dove	0	2

Using the same argument as before, the average payoffs to Hawk and Dove are $6 - 10p$ and $2 - 2p$ respectively, which are equal when $p = 1/2$. Once again, any mutant sub-population that tried to use Hawk and Dove in proportions other than 50:50 would be at a disadvantage as soon as it gained a toehold, and would be eliminated. Equal numbers of Hawks and Doves is an ESS.

But suppose Hawks were less aggressive, settling contests with less damage to each other, while still beating Doves at no cost. The payoffs might be:

	Hawk	Dove
Hawk	1	6
Dove	0	2

and now the average returns to Hawk and Dove are $6 - 5p$ and $2 - 2p$. But p is a proportion, and must be between zero and one, so the former *always* exceeds the latter. Hawks inevitably do better than Doves, so any Doves are eliminated, and the entire population uses Hawk all the time. No mutant using Dove at any frequency can invade—Hawk is an ESS.

By contrast, suppose Hawks were really brutal, and the average outcome of a contest between two Hawks was a payoff of -100 to each. Even here, Hawks are not driven to extinction, merely to a very low frequency. They survive at a low frequency because they meet few other Hawks, but still terrorize the Doves to gain fitness. Check that if you replace the '1' in the last table by -100, Hawks and Doves do equally well if the population plays Dove 25 times for every once it plays Hawk. Once again, no mutant using a different mixture can invade, so $(1/26, 25/26)$ is an ESS.

In any of these games with two strategies, there is an easy way to see whether there is some mixed strategy that is an ESS. Set out the table of payoffs as we have shown, and look down each of the two columns. There will be an ESS, provided that in the first column it is the second number that is higher, and in the second column it is the first number that is higher. That is, Dove does better against Hawk than Hawk does against Hawk; and Hawk does better against Dove than Dove v. Dove. When these conditions hold finding the ESS is immediate, and uses the same method as finding Colin's best mixed strategy against Roy in games such as that of Table 6.2. Subtract the numbers in the second row from those in the first. The first will always be negative, the second positive. Ignore the minus sign, and swop the answers round. This gives the two proportions. In the first game introduced, with payoffs:

	Hawk	Dove
Hawk	-2	10
Dove	0	3

this subtraction leads to $(-2, 7)$. Ignoring the minus sign and swapping, the proportions are $(7, 2)$, which corresponds, of course, to the answer we found. In the second table, the subtraction leads to $(-4, 4)$, and so in one more step to equal use of Hawk and Dove. If we followed the recipe for the third set of payoffs, the subtraction step gives $(1, 4)$; the first entry is not negative, so there is no mixed ESS.

These animals have no interest in some abstract notion of 'what is best for the population'. It would plainly be 'sensible' to settle all disputes by one toss of a fair coin, thereby removing all the losses through time-wasting posturing and mutual injury. But what matters in the struggle to survive is that you do better than your opponent, *even if*, overall, both of you do worse. Tennyson wrote of 'Nature red in tooth and claw'. But these models show that Hawk and Dove can co-exist, and indicate how changes in the value of resources, or the actions of the players, can lead to Hawk used more often, or less often. If Hawks could settle their contests in ways that led to less damage, they would expect to do better.

Such evolution does occur. Rather than immediately fight, animals might size each other up, through showing off their muscles, or demonstrating their lung capacity with a bellow. This is not Dove–Dove posturing; they really are prepared to fight. However, there are advantages to both animals if they can avoid doing so if it is plain that the contest would be one-sided. An excellent example is the study by Tim Clutton-Brock and his colleagues of the behaviour of red deer. In the mating season, a stag attempts to hold possession of a harem of hinds—success will lead to more offspring. Stags younger than seven years are generally not powerful enough to hold a harem, and fighting ability declines after age 11 or so. But between seven and 11 years, a stag has a reasonable expectation of mating. Stags who dispute a harem do not automatically escalate. Usually, they approach to about 100 metres, and alternately roar at each other; it seems they are attempting to judge who is the stronger. *Loud roars cannot be faked.* If this contest is inconclusive, they often move to within a few metres, and engage in a series of parallel walks, up and down, waving their ferocious antlers, seemingly seeking to persuade the other of the futility of a fight. Observations show that the longer the stags take over these preliminaries, the more likely they are to end up actually fighting. A substantial difference in size would be quickly detected by the roars or the walk; if these are indecisive, it is likely they are equally matched, and serious injury is possible.

The theme of this chapter is games with *few* choices. If there are just two tactics, there is always an ESS; sometimes it is a mixture of the two tactics, at other times there is an ESS when the whole population uses one or other of the two choices. But if there are three or more tactics, there are many more possibilities. Sometimes, there is an ESS that uses all the tactics on offer, but at other times there is no ESS at all.

The battle of the sexes

Both males and females can measure their reproductive success in terms of how many of their offspring reach maturity, capable of passing parental genes on to the next generation. Having offspring who do not themselves have offspring is a genetic waste. Richard Dawkins offered a simple model in which the two sexes have different strategies. In their efforts to succeed in this game, females can choose between being Coy or Fast, while males opt to be Faithful or a Philanderer.

A Coy female will refuse to mate until she is convinced that the male will share in the offspring's upbringing. Her idea is to have fewer offspring than she is biologically capable of, but to hope that more will reach maturity through the double input into rearing the brood. A disadvantage of her tactics is that she may drive away a succession of suitors not willing to dance to her demands for food, nest-building and so on: she may be left on the shelf.

A Fast female will copulate readily, perhaps with a stream of mates, and have many offspring. She may even find that one of her suitors is a Faithful male who shares the domestic chores, but that would be a bonus: she hopes that a good proportion of her offspring survive.

The units in a payoff table relate to the expected number of offspring *that reach maturity*, or even the expected number of grandchildren. Suppose any birth is worth 15 units to each parent, but it costs a total of 20 units to raise one child to maturity. If males are Faithful, that cost is shared equally. A Faithful male and a Coy female each lose the opportunity for offspring during their courtship, a cost of three units each, say. A Philandering male and a Coy female get nowhere, but waste no courtship time. The table of average payoffs to a female will be:

	Faithful male	Philanderer
Coy	2	0
Fast	5	−5

Suppose the proportion of Faithful males in the population is x. If a female is Coy, her average payoff is $2x$; and the average payoff to a Fast female is $5x - 5(1 - x) = 10x - 5$. These are equal when $x = 5/8$. So if the proportion of Faithful males exceeds 5/8, it pays to be Fast, while if it is less than 5/8, females are better off being Coy.

When males are Faithful, they prefer their mate to be Coy, as they have no interest at all in helping raise someone else's offspring. (This is not quite true: since you share genes with your near relatives, it is in your interest to help your brother's offspring to mature. Haldane famously pointed out that he would be prepared to sacrifice his life for that of two brothers, or eight first cousins. We skip over that glitch.) But should males be Philanderers, Coy females are very bad news. The male's table of payoffs will be:

	Coy female	Fast
Faithful	2	5
Philanderer	0	15

Male actions are dictated by y, the proportion of Coy females. For Faithful males, the average payoff is $2y + 5(1-y) = 5 - 3y$; a Philanderer gets $15 - 15y$, and the crossover point is when $y = 5/6$. With fewer Coy females, Philanderers are better off, but when there are more, Faithfulness is favoured.

The scene is set for a permanent merry-go-round, showing how the values of x and y can be expected to vary over time. If ever 5/8 of the males are Faithful, and 5/6 of the females are Coy, at the point marked X in Figure 6.3, there is no pressure to move away. But suppose there are surpluses of both Faithful males and Coy females, somewhere in the rectangle marked A. With so many Faithful males, Fast females are favoured, as they are able to bear many offspring and have someone share the burden of raising them. The genes that favour Fast females will now increase, so the population composition will tend to move into rectangle B.

In the region B, there are few Coy females, but many Faithful males. Philandering males now have the advantage, as nearly all the females are

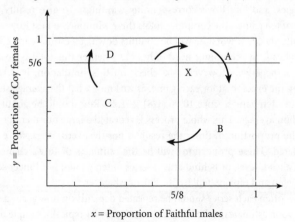

x = Proportion of Faithful males

Fig. 6.3 How the population composition cycles.

Fast, so Philanderers have many offspring, and their genes build up in frequency, moving the population composition into region C. Within C, Philandering males are in excess. Fast females normally have to raise their offspring alone, and so the Coy females will increase in numbers because they always have a Faithful partner to help in the upbringing. The pressure is to move the population composition into region D. And here, with many Coy females, the Philanderers do not get a look-in, the few Faithful males increase in numbers, and the population composition returns to region A to begin this cycle again.

There is no single best strategy for either sex; what you should do depends on how the opposite sex behaves (what a surprise!). A female prefers a Faithful male, as the first column of her payoff table dominates the second, and similarly males prefer Fast females. Male behaviour is at the behest of female taste—it was ever thus!

Postscript

Table 6.5 gives the probabilities that Roy will be ahead by some amount after a series of plays in the game of Table 6.2. It came from simulation, and not from direct calculation. Further details, and an analogy with opinion polls, may be useful. My computer has a built-in random number generator that can be used to choose the plays. In each game, the computer makes the choices of H or T for each player, according to the chosen strategies, and logs Roy's corresponding winnings. To give results for a series of ten games, the computer totals these winnings; at best Roy will be 20 units ahead, at worst he will be 30 units down. To estimate the chance he is ahead, the computer need only note whether this total is positive, zero, or negative. However, as a check on this simulation, it is worth storing the exact total for each game, as we know what the average should be. After ten games, since $10 \times (1/8) = 1.25$, Roy should be about 1.25 ahead on average. This whole process is repeated a large number of times, and the proportion of times the result is positive, zero, or negative then calculated. These proportions will be the estimates of Roy's chances of being ahead, level, or behind after a series of ten games. For longer series, start again and replace 'ten' by 50, 200, or whatever.

How often each series should be repeated depends on how accurate you want your estimates to be. Political opinion polls typically sample about 1500 electors, and warn that the estimates are subject to 'sampling error'.

This phrase refers to the fact that different random samples of 1500 voters will not contain exactly the same proportion giving allegiance to a particular party. (There are other reasons why an opinion poll may give an inaccurate answer, such as people not responding or telling untruths, the sample being unrepresentative of unmarried women, or retired men, etc., that have no direct parallel with our computer simulation.) Our estimates are indeed subject to sampling error, but this reduces as we increase the sample size.

The direct cost of an estimate by these methods has two components: first, a single cost to set up the 'experiment', and then a sampling cost that increases with the size of the sample. In opinion polls, the cost related to the sample size will be relatively high, as trained interviewers are paid according to the amount of work they do. In computer calculations, this sampling cost is often trivial, being merely the electricity consumed as the programme runs. The major cost in simulation is writing the computer programme, after which a sample of a million may cost little more than one of a thousand. High accuracy can be achieved by this method. Simulation is also very useful as a check on calculations in any probability problem. When we have calculated that a certain probability is, say, 24%, we might be able to simulate the process behind the calculation quite easily, and verify that we are not in gross error, at least.

Even when an opinion poll can avoid any sampling bias, it is still subject to the inevitable sampling error. When the true support for a political party is about 40%, a sample of size 1500 will be within one percentage point in nearly two polls out of three. But about one poll in 20 will have an error of at least 2.5 percentage points: in such polls, since the estimate for one party is too high, the estimate for its main rival will correspondingly be too low. Thus the apparent *difference* in their standings will be twice as wrong as the estimate of support.

Opinion poll organizations could increase their accuracy by taking a larger sample. But there is a severe law of diminishing returns: to be *twice* as accurate, the sample size must be *four times* as big. This will cost real money—much more than on a few kilowatt hours of electricity to power a computer through an enormous simulation.

Summary

Many problems can be set up as 'games with few choices'. Some are obvious ones that arise naturally as games, such as matching coins or showing counters, and the table of payoffs can easily be written down.

Others, including social conflicts (the dating game, the guilty husband) need some thought in allocating suitable payoffs, and the two players may have very different ideas of what these should be. No matter: set up your own payoff table, from your standpoint, and seek to maximize your reward. In all our games, we have assumed the players make their choices without knowledge of what the opponent has selected.

Your first step should be to eliminate any dominated strategies. This simplifies the table. Do not worry if you overlook something you should have struck out—these methods will eventually lead to it being allocated zero probability, but you will have more possibilities to search through. Then look for a saddle point: if you find one, the game is solved and the play is predictable.

If each player has at least two choices, and there is no saddle point, then optimal play by both sides leads to mixed strategies. If both players have three or more choices, I have not described how to proceed—you must look elsewhere. (There will be a solution, and there are algorithms to obtain it.) We have seen how to solve games in which one or both players have only two choices.

The easiest case is when each player has exactly two choices. The three-step method of subtracting one column (or row) from the other, swapping the order and discarding the minus signs gives the proportions of time for each strategy. The value of the game is found, and the overall game is made fair if Roy pays this amount to Colin each time.

If Roy has two choices, and Colin has more than two, the solution follows by looking at the series of 2×2 games, in which Colin deliberately restricts himself to just two of his choices. Find Roy's best strategy in each of these games; having found it, now work out the *worst* possible outcome to Roy, whichever choice Colin makes. This is the amount Roy can guarantee to win, using that strategy. Find the *biggest* of these worst outcomes. That is the value of the game, and that also indicates Roy's best strategy. The diagrammatic method will often be the most convenient to use.

Suppose Roy is the one with many choices, while Colin has just two. One way to proceed is to exchange the roles. However, do not forget that the payoff table has been drawn up from Roy's standpoint, and his interests are the opposite of Colin's. To deal with this change, change the sign of all the original payoffs: a payoff of $+2$ to Roy becomes one of -2 to Colin, while -3 to Roy is equivalent to $+3$ to Colin. The payoff table now has the format we are used to. The rows player seeks to maximize his payoff. We know how to advise him.

It is rather ironic that if one player goes to the trouble of working out his optimal strategy in a 2 × 2 game, and using it, it is quite irrelevant what the other player does, in the long run: the average payoff is the same. The lazy or incompetent player is protected from his own inadequacies by the good play of the other! In the game of Table 6.2, so long as Roy shows H 3/8 of the time, Colin will lose 1/8 on average, whatever he does. But if he realizes this and stops paying attention, simply showing H every time out of idleness, Roy could take advantage, and show T even more frequently. So Colin does, after all, have an incentive to use his best strategy.

Suppose Roy has two choices, and Colin has three or more. Then Roy's best strategy is a mixed one, and Colin has a best strategy that mixes just two choices; so long as Roy uses his best strategy, Colin can use the choices in his own best strategy at *any* frequency—the average payoff is the same. But now there is an extra danger to Colin: if he slips in some plays using the *other* choices, he will usually be giving money away. Best play by Roy no longer protects Colin from his own poor play.

This last point is even more relevant when each player has three or more choices. It will often be the case that some of these choices should never be used, so it is vital to discover which they are; using them against a good player is a sure-fire guarantee of losses.

Game theory has proved useful in models of animal behaviour. Animals will typically be in competition for a resource that would increase their reproductive success. When there are just two tactics available, and both can co-exist, there is a single definite population composition that will resist all attempts to invade it. This evolutionarily stable strategy, or ESS, pays no attention to whether the resources are allocated in some 'efficient' manner; it simply secures the population against invasion by a mutant using a different mixture of the two tactics. With three or more tactics, there may be no ESS at all.

Test yourself quiz

6.1 For the table shown, find the maximin and the minimax. Is there a saddle point? Would there be one if the entry '15' were replaced by '6'?

	A	B	C
I	12	4	15
II	8	9	7
III	4	10	3

6.2 Carry out the computations for Table 6.6 in the text, and verify that the best strategies for Roy and Colin are those stated.

6.3 For the Table in Q6.1 above, is there a dominated strategy for either Rows or Columns? What could you replace the entry '10' by to make III a dominated strategy that Rows would therefore never use?

6.4 If the strategies A and III were not permitted in this game, so that Rows used only I or II, and Columns had only B or C, what are the best tactics for Rows? For Columns? How much should Rows pay Columns to make the game fair?

6.5 Describe a practical way of using an ordinary fair die to choose the numbers 1 to 10, each with probability 1/10.

6.6 Each of Roy and Colin selects any number of coins from zero to four; the winner is decided on the basis of the *total* number of coins between them. Roy wins if the total is zero, one, two, six, seven, or eight; Colin wins if the total is three, four, or five. The loser pays £1 to the winner.

Draw up the 5 × 5 payoff table. Look at it from Roy's perspective, and notice you can eliminate some dominated strategies; do so. Look at what is left from Colin's perspective, and hence throw out some of his options on the same basis. Continue as far as you can; you *ought* to get it down to a 2 × 2 table—solve this game. What are the best strategies for the two players? Is it a fair game?

6.7 Change the previous game so that Colin is the winner, not Roy, if the total is two or six. What difference does this make?

6.8 In a Hawk–Dove game, suppose the value of the prize is 12 units; Hawks who fight each suffer eight units of damage; Doves who posture waste three units until they retire (or win). Draw up the payoff table, and decide what the stable population composition will be (the ESS).

6.9 Consider what happens to the Hawk–Dove game of the previous question if the value of the prize is altered while everything else stays the same. First look at prizes higher than 12, then lower. Check that your answers are sensible.

CHAPTER 7

Waiting, waiting, waiting

The common theme here is that there is a sequence of opportunities for something to happen. In some problems, we want to know how long we have to wait, in others it is how many 'somethings' are going to occur. As each problem is introduced, you are invited to make your assessment of the appropriate answer. Be honest: pause before you read on, and make your best guess. Award yourself gold stars for good guesses—provided your reasoning was valid—and suffer some suitable penance otherwise. Do not simply read on to discover the answer and then persuade yourself you would have got it about right. Be active.

Birthdays

How large should a group of people be, to make it more likely than not that at least two of them share a birthday? To begin to answer this, we have to know how the group has been chosen. Were they all volunteers in a psychology experiment, based on comparisons of identical twins? Is this a gathering of a group of Aquarians, born between 21 January and 19 February, seeking to compare some astrologer's predictions? In either case, there would be good reason to expect a pair with a common birthday. But when this question is posed, it is usually implicit that the group of people have been drawn together in some fairly random fashion, certainly unrelated to their birth dates. This assumption is made here.

Rather more people are born in the summer than in the winter. Only leap years contain a 29 February. Taking complete and exact account of these glitches is rather complicated. The mathematician's motto is that if a problem looks hard, replace it by an easier one, and then try to modify the solution. So here we first assume that birthdays can arise only on 365 days, ignoring 29 February, and then suppose that all these dates are equally likely. If we cannot solve this simpler problem, we have no hope with the original, more complex one; but if we are successful, its solution may point the way forward.

Trick 1 for finding the probability that something happens is to calculate the chance it does *not* happen. That is the method here. Either all birthdays are different, or at least two of them coincide. Find the chance that everybody in the group has *different* birthdays, and then subtract this from 1 to obtain the probability that at least one pair have the same birthday.

With just two people, the second person has a different birthday 364 times out of 365. A group of three will all have different birthdays when the first two are different (this happens with probability 364/365), and then the third person's birthday is on one of the remaining 363 days. This means that the total chance all three have different birthdays is

$$\frac{364}{365} \times \frac{363}{365}.$$

For a group of four, all the first three must have different birthdays as above, and the fourth person's birthday must be on any one of the other 362 days. So the chance all four have different birthdays is

$$\frac{364}{365} \times \frac{363}{365} \times \frac{362}{365}.$$

The pattern should be clear by now. The chance that all in a group of five have different birthdays is

$$\frac{364}{365} \times \frac{363}{365} \times \frac{362}{365} \times \frac{361}{365}. \tag{*}$$

For larger groups, we multiply successively by 360/365, 359/365, and so on.

It is convenient to have a shorthand way to express (*). Write a product such as $364 \times 363 \times 362 \times 361$ as $(364)_4$. (To ensure you can handle this notation, check that the statements $(6)_2 = 30$, $(4)_5 = 0$ and $(5)_3 = 60$ make sense to you.) We can now write the chance that five people all have different birthdays as

$$\frac{(364)_4}{365^4},$$

and then, for example, the chance that ten all have different birthdays will be

$$\frac{(364)_9}{365^9}.$$

For small groups, these probabilities are close to 1. They decrease as the group size increases, since each is found from its predecessor via multiplication by a fraction that is less than 1. If we pursued this all the way to a group of size 366, the probabilities that all birthdays are then different would be zero. We seek to find where the probability that all are different crosses over from above 50% to below 50%. That is where the odds change from favouring all different birthdays to favouring at least two with the same birthday.

When people meet this problem for the first time, and are asked for a speculative guess as to the smallest size of the group that makes the odds *favour* a pair with a common birthday, most offers are wildly wrong. The value 183 is fairly popular, for the obvious reason that it is just more than one-half of 365. It is quite rare for initial guesses to be below 40. And yet the answer is 23.

One way to convince you of this answer is that you use a pocket calculator to do these sums. First calculate the chance that all in a group of size 22 have different birthdays, then do the same for a group of size 23. These respective chances are

$$\frac{(364)_{21}}{365^{21}} \text{ and } \frac{(364)_{22}}{365^{22}},$$

which evaluate to 0.5243.. and 0.4927.., so 23 is indeed the crossover point. Once this arithmetic persuades you that this is the correct answer, I can show a less exact analysis, but one that may make this result less surprising.

For any pair of people, the chance they *do* have the same birthday is 1/365. This is quite tiny, but a large number of tiny chances can amount to something substantial. ('Large oaks from little acorns grow.') See what happens as the group builds up in size. If the first ten people contain a pair with the same birthday, then any larger group will do so. So assume these ten all have different birthdays. When the 11th person joins, s/he has ten chances, each of size 1/365, of having a common birthday with someone already present. If there is no common birthday, the next person has 11 chances, the one after has 12 chances, and so on. The total number of chances grows quite quickly. In a group of 23 people, there are (Appendix I) $^{23}C_2$ *pairs* of people, which works out to be 253. A group with 23 people generates 253 chances, each of size 1/365, of having two with the same birthday. Admittedly, these chances have some overlap: if A and B have a common birthday, and also A and C have a common birthday, then inevitably so do B and C. But the realization that a group of only 23 people generates as many as 253 chances may help you appreciate why, taken together, the overall chance of a pair with matching birthdays reaches one-half.

How are these conclusions affected if we move closer to the real situation by accepting that not all days yield the same number of birthdays? The central fact is that any move *away* from a uniform distribution of birthdays over all 365 days tends to *increase* the chance that a group of a given size has a pair with the same birthday. (Perhaps surprisingly, no 'advanced' mathematics is needed to show this. The essence of the proof is outlined in the boxed material.)

Even when all dates in a 365-day year are equally likely, a group of 23 or more people is more likely than not to contain a pair with matching birthdays. So the same conclusion certainly holds when we bring in the non-uniform distribution of birth dates. Depending on how uneven this

---

A non-uniform distribution increases the chance of coincident birthdays (outline argument)

Suppose not all days in the year are equally likely to be birthdays. Then some dates have different probabilities from others—choose two of these dates, and suppose their respective probabilities are x and y, with x above average, and y below average. Whatever the size of the group, there is some gigantic expression, involving the probabilities attached to all 365 days, that everyone in the group has a different birthday. (For a group of size four, there are over 700 million terms!) Call the value of this A.

Now alter matters by replacing both the probabilities x and y by their average, $(x + y)/2$. *Keep all the other probabilities the same*, and write down the new expression that everyone in the group has a different birthday. Call the new value B. Only two of the 365 probabilities have been changed, so although the expressions for A and B are lengthy, most of the terms in them are identical. So when we work out the difference $B - A$, the vast majority of terms cancel out. When you write down the detail, this difference $B - A$ collapses down to $(x - y)^2/4$, multiplied by something that is known to be positive, *because it is a sum of probabilities*.

Since $(x - y)^2/4$ is a square quantity, it is also positive, so $B - A$ is positive, i.e. B is bigger than A. This means that our alteration has *increased* the chance that all the birthdays are different. And replacing both x and y by their average plainly makes the probabilities *more uniform*.

As making the probabilities more uniform increases the chance that all the birthdays are different, making them less uniform must increase the chance that at least two of them are the same.

distribution is, it may even be more likely than not that a random group of only 22 people has a pair with a common birthday!

How to take account of 29 February? Begin by supposing all 366 days in a leap year are equally likely. The corresponding calculations for groups of sizes 22 and 23 all born in a leap year show that the chances that they consist of people all with different birthdays are 0.5254.. and 0.4937.. respectively. Once again, the crossover point between more and less likely is as the group increases from 22 to 23. If there were 366 equally likely birth dates, it is still more likely than not that a group of 23 people have a pair with a common birthday. Because 29 February has a smaller chance than its neighbouring dates of being someone's birthday, the true distribution is not uniform across all 366 possible dates. This non-uniformity always tends to *increase* the chance of there being at least one coincident birthday.

Summarizing our findings: a group of 23 or more randomly selected people is definitely more likely than not to contain a pair with the same birthday. Depending on how non-uniformly across the year birthdays occur, groups of size 22 may even be large enough for the odds to favour a pair of matching birthdays.

The science writer Robert Matthews put this statement to a statistical test. His groups of 23 people were the 22 footballers and the referee who were on the soccer field at the beginning of the ten Premier League matches played on 19 April 1997. Six of these games had some pair with the same birthday, and all 23 men had different birthdays in the other four games. It would be unreasonable to demand better agreement between data and theory. (November and December birthdays were more common than average.)

Other places where this theory can be tested against real data are in schools, where many class sizes are in the range 26 to 32. With birthdays assumed randomly but evenly distributed over 365 days, the probabilities that all birthdays are different is 40% in groups of size 26, and 25% in groups of size 32. So about 60% of classes with 26 pupils, and 75% of classes with 32 pupils, *will* have a pair with a common birthday. Gather together two classes of size 25 to make one 'class' of size 50: the chance of a pair with the same birthday exceeds 97%.

The Spoilers (2)

In the last chapter, I noted one use of probability in this novel by Desmond Bagley. There is a second episode. Two men have a potentially long and boring car journey in a sparsely populated area. The numerate driver, Follett, offers his passenger, Warren, a series of bets to while away the journey. They will focus on the *last two digits* in the car number plates of all the cars they see travelling in the opposite direction. So both J796 OPB and LAP 296C would yield '96', B28 SWW would give '28' and so on. There are 100 different outcomes possible, running from '00' to '99', and both men are content that it is reasonable to assume a random and uniform distribution over this set of possibilities. Follett offers to bet, at even money, that at least two of the next 20 cars would yield the same outcome.

Warren accepts this bet. He reasons that with 100 numbers, the chance that any two are the same is 1/100, so with 20 cars the chance is 20/100—the odds are in his favour. Poor sap. He loses almost every bet. At the end of the journey, as he collects his winnings, Follett points out Warren's fallacy.

This problem is identical in format, but with different numbers, to the birthday-matching problem. There are 100, instead of 365, possibilities, and we want to know how likely it is that all in a group of a given size are different. Taking all 100 outcomes as equally likely, the chances that they are all different in a batch of 20 cars is

$$\frac{(99)_{19}}{100^{19}},$$

which evaluates to a miserly 0.1304. Follett expected to lose about 13% of all bets, and win the other 87%. He could have offered odds of five to one, and still the game would have favoured him, but why should he do this when Warren was prepared to play at even money?

Twenty cars gives a heavy bias in favour of a coincident pair of numbers. What number of cars would give a fair game? Such a game might give children some distraction on a long car journey, and subtly instil some notions of probability. Press the right buttons on a pocket calculator to see that with 11 cars, the chance all numbers are different is 0.565; with 12 it is 0.503; with 13 it is 0.443. So batches of *12* cars give a virtually fair game: 50.3% of the time, all 12 numbers will be different, 49.7% of the time they will include a coincident pair.

Other matches

Samples of size 23 when there are 365 possibilities, or 12 with 100 possibilities, give roughly equal chances of a match, or of no match. It would be convenient to have a simple formula that identifies these crossover points when there are N different possibilities. It turns out that you will not go far wrong if you take the crossover point as being at $\sqrt{1.4N}$. When $N = 365$, this formula gives the answer $\sqrt{511} = 22.6$, well in accord with the true value 23. When $N = 100$, the formula leads to $\sqrt{140} = 11.8$, plainly consistent with the 12 we have just seen. To see where this formula comes from, read the boxed analysis.

In the UK National Lottery, there are nearly 14 million possible combinations. Using our formula with that value of N shows that about 4400 draws are required to give an evens chance of containing a repeat of a jackpot-winning combination. At two draws per week, this corresponds to

---

How many for an evens chance of a match?

Suppose there are N different possibilities to choose from, all equally likely. When we select a pair of them at random, the probability they are the same is $1/N$. So the probability they are different is $1 - 1/N$. In a sample of size K, there are $^{K}C_2 = K(K-1)/2$ different *pairs*. (We used this argument earlier to obtain the value 253 when $K = 23$.) These pairs are not all independent, but many of them are independent of many others: if I am told whether A and B are the same, and also about C and D, this says nothing about A and C. So let's act as though all the different pairs were independent, accepting that this is not quite right.

If they all were independent, the chances that all the pairs are different would be

$$(1 - 1/N)^M, \text{ where } M = K(K-1)/2.$$

When N is large, it is a mathematical fact that the value of $(1 - 1/N)^M$ is well approximated by $e^{-M/N}$. So the chances of a match will be about evens when this quantity is near $1/2$. Take natural logarithms to find this corresponds to M/N being about $\ln(2)$. But since $M = K(K-1)/2$, this leads to

$$K(K-1) \approx 2N\ln(2).$$

Push the approximations a little further. $K(K-1)$ is a little less than K^2, and $2\ln(2)$ is a little less than 1.4, so we change this last expression to

$$K^2 \approx 1.4N.$$

Take square roots of both sides to find the result stated: the size K that gives an evens chance of a match is approximately $\sqrt{1.4N}$.

Table 7.1 Assume there are N possible outcomes, all equally likely. For the different odds shown in the first column, the second column gives the approximate sample size required for at least one repeated outcome. Examples are shown in columns 3 and 4.

Approximate true odds	Sample size needed	e.g. birthdays with $N = 365$	Car numbers with $N = 100$
evens	$\sqrt{1.4N}$	23	12
2–1 on	$\sqrt{2.2N}$	28-29	15
3–1 on	$\sqrt{2.8N}$	32	17
6–1 on	$\sqrt{3.9N}$	38	20
10–1 on	$\sqrt{4.8N}$	42	22

about 42 years. So most people aged 30 or under when the Lottery began have a reasonable chance of seeing a previous jackpot combination triumph again in their lifetime, assuming the 6/49 format survives. But *which* combination? Your guess is exactly as good as mine!

(This same notion could be used by Camelot's statisticians to put the Lucky Dip random number generators to the test. If Lucky Dip genuinely selects numbers with no reference to previous outcomes, then a batch of about 4400 Lucky Dip selections should have a 50% chance of containing an identical pair.)

As well as looking for bets at even chances, there are similar formulae that act as excellent guides for fair bets at other sets of odds. Any use you make of Table 7.1 to lever the odds comfortably in your favour, much as Follett did to Warren, is entirely a matter for your own conscience.

More than one match?

In Robert Matthews' data on soccer players' birthdays, he found that among his ten soccer matches, there were two in which there were two pairs of participants having common birthdays. Was that unusual? An exact analysis is possible, but a method similar to that in the box above gives an excellent approximation, with much less effort. There are good statistical reasons—a large number of fairly independent small chances—why the actual number of matching pairs will tend to conform to a *Poisson distribution* (Appendix II). We find the *average* number of matches in the group of people, and then use the corresponding Poisson distribution to calculate the chances of exactly 0, 1, 2, 3, . . . matches.

We know that a group of size K contains ${}^K C_2 = K(K-1)/2$ pairs. Each pair has chance $1/N$ of being matched (e.g. same birthday), so the average number of matches is $K(K-1)/(2N)$—call this value μ. For the birthdays example in which $N = 365$ and $K = 23$, the value of μ is 0.693. Using the formula in Appendix II, we find that, on average, there would be five of the ten groups with no coincident birthdays, three and a half with one pair, and one and a half with two or more pairs. In Matthews' actual data, there were four groups with no coincident birthdays, four with just one pair, and two with two pairs. The data match the average values very closely. His figures are not at all unusual, they are just about what we would expect.

What about the chances of a triple, or a quadruple, match? How large should a group be for such an event to have a reasonable chance? Similar arguments will give estimates of the necessary sizes. With the same notation as in the boxed argument, consider a triple match in a group of size K. In a year with N equally likely birth dates, the *average* number of triple matches is

$$\frac{K(K-1)(K-2)}{6N^2}.$$

This leads to the estimate

$$K \approx 1.6N^{2/3}$$

as the approximate size of a group that has a 50% chance of containing a triple match. When $N = 365$, this shows that a group of about 82 randomly selected people has a 50% chance of containing three people with the same birthday. With car number plates modified as before to fall in the range 00 to 99, there is a 50% chance of a triple match in the next 35 cars that you see.

For a quadruple match, the corresponding value is $K \approx 2N^{3/4}$. For birthdays, this gives a value of 167. Take a random selection of 167 members of parliament; there is a 50% chance that there is a group (gang?) of four having the same birthday.

Packs of cards (1)

Shuffle two packs of ordinary playing cards separately, and place them face down in front of you. Turn over the top cards in each pile, and compare them to see if they match exactly. Perhaps you show the Nine of Clubs, and the Six of Hearts—a non-match. Move on to the second cards, and maybe turn over the Eight of Hearts and the King of Spades—another non-match. Work your way through all 52 cards in each pile, and simply count the number of times the two corresponding cards match. You might find none at all; you could have one match, two, three, and so on. About how many matches would *you* expect, on average? Make your guess now: the truth will emerge shortly.

When I describe this problem to people, a common guess is about four or five. Very few guess more than this, not many suggest fewer than two matches. But to see what the average number is, consider the experiment one pair of cards at a time. Whatever card is at the top of the left pack, there is a 1/52 chance that the same card is at the top of the right pack. So the chance the top cards match is 1/52. Move on to consider the second cards in each pile. *Do not be distracted by any red herrings about whether or not the top cards have matched.* If you were told what the top cards were, that would indeed affect the chance that the second cards match, but simply look at the second cards in isolation. For the same reason as with the top cards, the chance the second cards match is also 1/52. The same applies to any specific position, fifth, twenty-fourth or whatever—the chance of a match is 1/52. And this leads straight to the answer: there are 52 chances of a match, each of size 1/52, so the average number of matches is $52 \times (1/52) = 1$.

A moment's thought assures you that a parallel argument would apply to packs of any size, given only that each is identically composed with distinct cards. If you extract all the Hearts and all the Spades, shuffle each suit separately and perform the same experiment, the average number of matching ranks is $13 \times (1/13) = 1$. Take two sets of pool balls numbered 1 to 15, mix them thoroughly in separate bags, withdraw one ball simultaneously from each, and compare their numbers. On average, there is one match as you draw out all the balls.

The simplicity of this argument is disarming. People may fail to see the correct answer immediately perhaps because they appreciate that the *distribution* of the total number of matches is quite complicated. This complication obscures the simplicity of the argument based on averages. Essentially, we used the idea, explored further in Appendix III, that *the average of a sum is the sum of the averages of its components*. Averages often reach the heart of the matter immediately.

Although the exact distribution of the number of matches has a fiendishly complex form, our friend the Poisson distribution gives an excellent approximation, whenever our 'packs' have at least (say) eight 'cards'. The average number of matches is always one, so we can use the formula in Appendix II to show the main values for the probabilities in the table.

Number of matches	0	1	2	3	4	5 or more
Probability (%)	37	37	18	6	1.5	0.4

You can use this table to derive fair odds for bets such as 'There will be at least x matches', for different x.

We can dress this problem up in different colours, far removed from packs of cards. Here are three scenarios—you can think of others.

- N letters have been written to individuals, and there are N corresponding envelopes with names and addresses. The boss's scatterbrained secretary allocates the letters at random to the envelopes. On average, one letter goes to its intended recipient. (A variant on this has a hat-check porter returning headgear at random after the opera.)

- M young couples are together for a hedonistic fortnight on a Greek island. The combination of sun, sea, ouzo, etc. leads to random re-alignments. On average, just one original couple are together at the end of the stay.

- John arranges his books on shelves alphabetically by authors' names, and within the same author by year of publication. The books are removed so that the room can be redecorated. He replaces the books, but this time alphabetically by title. On average, just one book is in its original place—no matter how large his library.

Packs of cards (2)

This demonstration can appear quite amazing, but it will occasionally end in disaster. Shuffle an ordinary pack of cards, and call for a volunteer—Sarah—from your audience. Explain openly to her and the audience what you intend to do. You will deal the cards, face up, at a steady pace. Sarah will secretly select any one of the first ten cards, giving you no clues as to which, and start counting from that card. Suppose she selects the fourth card, the Three of Diamonds. This Three means that she (silently) counts on three cards as you continue to deal; perhaps this third card is the Four of Hearts (see Figure 7.1). She now counts on four further cards, reaching

Fig. 7.1 The cards that Sarah sees.

(say) the Ten of Diamonds, and so on. Any Ten, Jack, Queen or King is to be regarded as worth ten in her counting, while an Ace just counts one.

The last card she hits must be somewhere within the bottom ten cards in the pack. For example, if she hits the 48th card, a Seven, that will be the last card. But if the 48th card is a Two, she counts on to the 50th, the Queen of Clubs, say, and that is the last card she hits. Your object is to guess which is this last card she hits. Estimate your chance of success.

To begin with, you might think she is equally likely to hit any of the last ten cards, so your chance of guessing right is one in ten. But in the example just given, where card 48 is a Two, it is impossible for that card to be the last one she hits. Similarly, if card 45 is an Ace, and card 47 is a Four, she cannot have either of those as her last hit. Your chance of guessing right is going to be rather better than one in ten—maybe it is up to one in six, or one in seven. I claim that the use of skill can increase your winning chance to over 70%.

What skill can you employ to achieve such a success rate? As you deal out the cards, do not attempt to read Sarah's lips or mind, but concentrate on the cards. You are yourself doing exactly what you asked her to do: you select one of the first ten cards, and count on according to the recipe given, to see where you end within the last ten cards; that card will be your guess as to Sarah's endpoint.

Why should this work? Since you have no idea where she began, your chance of starting at the same card as she chose is one in ten. If you are lucky enough to have chosen the same card as she did, the two of you will be locked together as you both silently count on, and you will inevitably hit the same last card. Nine times out of ten, you will choose a different start point. However, wherever you begin, you will have several chances of hitting one of the cards in her sequence, and if you ever do so, the two of you will remain locked together to the finish. Your 'guess' is then correct.

We can estimate your chance of locking in to her sequence. On average, her initial selection is around card five. There are four ranks (Ten, Jack, Queen and King) that move her on ten cards, so the average distance moved in one step is between six and seven cards. Over 52 cards, on average, Sarah will hit about eight cards altogether, including the first one. When *you* are choosing where to begin, you should deliberately select a low card: if the first few cards are Seven, Ten, Four, Queen, King, Two, ... you will already have begun with the Four, as this increases your opportunities to land on a member of her sequence. Roughly speaking, since you also jump about six or seven at a step, your chance of *missing* any given card is about 6/7; there are about eight cards that Sarah hits, so your chance of missing all of them is about $(6/7)^8 = 0.29$. That would give you a chance of about 71% of hitting at least once—and one hit suffices to lock you in to her sequence. *You only have to be lucky once.*

That argument is a bit crude, but it gives a ball-park estimate. An exact calculation would need to consider every possible arrangement of the 52 cards, which is not practicable. Far better is to experiment either by dealing out a shuffled pack a few times, or by computer simulation. The latter is more reliable, and promises the chance of simulating very large numbers of cases to get good estimates. For each 'shuffle', look at each of the first ten cards as possible start points, and follow through to see which one of the bottom ten cards you finally hit. Sometimes, whichever of the first ten cards you begin with, you inevitably hit the same last card. In this case you are certain to lock in to Sarah's sequence, wherever you both begin. At other times, the ten different start points lead to more than one possible finishing card, but it is quite straightforward to calculate the chance that you and Sarah reach the same final card. For example, maybe seven start cards lead to card 46, and the other three start cards have sequences ending up at card 51. Both of you will reach the same last card whenever you both begin with one of the block of seven start points, or when you both begin from among the block of three. The total chance is $0.7 \times 0.7 + 0.3 \times 0.3 = 58\%$. The calculations are just as easy when there are three or more different finishing cards possible. In my simulation of 10000 shuffles, nearly 2800 gave the same last card, for all ten permitted start points. Assuming you and Sarah both choose your start point at random, the overall estimate of the chance you both end at the same card is just over 70%. (It is something of a fluke that our crude 'estimate' of 71% is as close as this. Given how that estimate was made, any true answer between 60% and 80% would have been very satisfactory.)

You can expect to do a little better than this: your intention is to start at the first low card that turns up, and Sarah is likely to start her count by the sixth or seventh card. If both of you selected one of the first five cards at random, the average probability you will get locked together somewhere over the 52 cards now exceeds 75%.

All this relies on accuracy in counting. If either of you makes an error after you have locked in together, you may not recover before you run out of cards. So ensure that Sarah has a good concentration span and is competent at her task. You can help her by the even pace with which you deal out the cards. You might offer a simultaneous display, in the manner of a chess grandmaster: you will guess the end cards of six volunteers, who all choose their start cards independently. Your guess will be the same for all participants—you are banking on all of them, and you, becoming locked together. This is where your demonstration can go spectacularly wrong; your 'assistants' all end up on the Nine of Diamonds, card 48, while you are confidently offering card 50, the Six of Clubs. You might guard against this by doing the whole experiment with two packs of cards shuffled together. This gives you twice as many chances to lock in to the sequence(s) of the others, and you would be very unlucky not to do so as you deal out 104 cards. Your chance of success with a double pack is over 90%, rather more if there is a bias towards choosing early cards. There is an increased risk of random errors creeping into the count, so keep a steady pace as you deal.

Card/coupon collecting

Complete sets of pre-war cigarette cards are sought after by collectors. Packets of cigarettes would include a picture and potted biography of a well-known sportsman, and these collections give snapshots of golf, soccer or cricket around 1930. Today, boxes of cornflakes, or packets of children's sweets, may contain similar items. If you set out to acquire a complete collection, about how many purchases would you expect to make to achieve your goal?

As with birthdays, we must make some caveats. Have the manufacturers played fair by putting in equal numbers of all the items? We shall begin by assuming this, and also that the cards are randomly scattered without regional bias. Each purchase is taken to generate exactly one card, and we will ignore the possibilities of swaps. Suppose there are 24 different cards. Estimate how many purchases, on average, you would expect to make to obtain a complete collection. Separately, if there are twice as many cards, would you expect to have to make twice as many purchases? More than twice as many? Fewer than twice as many? The answer is not at all obvious.

Your first buy certainly gives you a starter for the collection. With 24 to aim for, you would be very unlucky if you found the same card in your second purchase—you get a different one 23 times out of 24. As you begin to build up your collection, you normally get a new card with each buy, and only occasionally are you disappointed. Perhaps your first 15 purchases give you 12 different cards, with just three repeats. Consider the next buy. As there are 24 cards altogether, and you have 12 of them, there are 12 new ones you still seek, so you have equal chances of a new card, or a repeat. *If an event has probability p of happening, it takes an average of 1/p goes for it to happen* (Appendix III). At this stage, $p = 1/2$, so $1/p = 2$; on average, we need two purchases, to get the 13th card, once we have acquired the 12th.

By the time you have collected 23 cards, you will surely have accumulated many repeats. There is one more card to be found, but each purchase has only a 1/24 chance of containing it. So, however many buys it took you to obtain the first 23 cards, you will need another 24 buys, on average, to get the last one.

In summary, new cards arrive very quickly at first, then about once in two buys when you have half your target, but you can expect long intervals between new cards at or near the end. To quantify this properly, good notation helps. Write $X(1)$, $X(2)$, $X(3)$, ... , $X(24)$ for the numbers of purchases you need for each of the cards in succession. $X(1)$ is known to be 1, we are disappointed if $X(2)$ is not 1, we have seen that $X(13)$ averages at 2, and that $X(24)$ averages at 24. The actual total number of cards is

$$X(1) + X(2) + X(3) + \ldots + X(24). \tag{*}$$

If we can find a formula for the average number of a typical value in this sum, then we can use the recipe that the *average of the sum is the sum of the averages* (Appendix III), and use (*) to obtain the average number of purchases altogether.

Take, as an example, $X(18)$. We will find its average, and then it will be fairly obvious what all the other ones must be. Recall that $X(18)$ is the number of purchases required, after we have got 17 different cards, to obtain the 18th. Since there are seven new cards to find, and 24 equally likely cards, the chance that any buy yields one of the new cards is 7/24. Recall the result we are using frequently: if an outcome has probability 7/24, the average time to get it is 24/7. So the average value of $X(18)$, the time to wait for one of seven cards, is 24/7. The pattern shouts loud and clear: the average for $X(19)$, the time to wait for one of six new cards, will be 24/6, and so on. Taking the average of each of the quantities in (*), and deliberately writing $X(1) = 1$ as 24/24, the *average* of (*) is

$$\frac{24}{24} + \frac{24}{23} + \frac{24}{22} + \dots + \frac{24}{7} + \dots + \frac{24}{2} + \frac{24}{1}.$$

Take out the common factor 24 from the top, and write the sum in reverse order: the average becomes written as

$$24\left(1 + \frac{1}{2} + \frac{1}{3} + \frac{1}{4} + \dots + \frac{1}{24}\right).$$

Its numerical value is about 91. This is the average number of buys to collect all 24 cards, if swaps are not permitted.

Plainly, the corresponding average for a set of 48 cards is

$$48\left(1 + \frac{1}{2} + \frac{1}{3} + \dots + \frac{1}{48}\right) = 214 \text{ or so.}$$

So a set that is twice as big will take *rather more* than twice as many purchases to obtain. A set with some general number, N, objects takes

$$N\left(1 + \frac{1}{2} + \frac{1}{3} + \dots + \frac{1}{N}\right)$$

purchases, on average, to obtain the full set. The sum in parentheses has an honoured place in mathematics, and it is handy to know that there is an excellent approximation to it, as working out its exact value can be tedious. Using this approximation (see advanced maths texts for details), we can compute the average number of purchases, for large N, as being close to

$$N\{\ln(N) + 0.577\}.$$

This is much easier to evaluate with a pocket calculator, and the answers differ by less than 2% from the exact answers displayed earlier, when N exceeds 10.

To see this formula in action in a different context, contemplate how many draws are needed in the UK National Lottery for all the numbers to appear at least once as the Bonus number. The last formula with $N = 49$ leads to the answer of about 219 draws, on average. In fact, the first 30 draws gave 26 different Bonus numbers, but then more repetitions began to appear. The next 20 draws produced ten new numbers, and by draw 77 only five numbers had not arisen at all. On average, after draw 77, the next new Bonus number would take $49/5 \approx 10$ draws. It actually took 36 more draws, and then another 16 for the 46th, and 26 more to get the 47th. At that stage, draw 155, with two numbers not yet chosen, the average wait for the first of them would be $49/2 = 24.5$ draws, and then an average of another 49 draws for the last one. The wait for the final number accounts for over 20% of the total. A hundred draws later, neither had appeared, but that last calculation, that a *further* $24.5 + 49$ draws, on average, would be needed, still stood. Then suddenly, the final two Bonus numbers arrived in consecutive draws, and the 'collection' was complete at draw 262—a little late, but not unusually far from the original estimate of 219 draws. (The calculation for the average time to wait until all the numbers have been drawn as main numbers is more complicated, as each draw generates six numbers. Similarly with modern collections of sportsmen or pop stars: children are encouraged to buy packets, each of which will contain five of the desired objects.)

Another application of these ideas to the National Lottery is in estimating how many tickets should be sold so that all of the 14 million different combinations have actually been bought for any draw. We saw in Chapter 2 that not all combinations have the same chance of being chosen: but just as any move away from uniformity tends to increase the chance of coincident birthdays, so if some combinations are more popular, that tends to increase the chance of *not* selecting all of them. Thus, if we calculate the average number of tickets that would be needed to cover all combinations if selections were made at random, we can expect this to underestimate the true number required. Using the above formula with N just under 14 million, we see that the sales would have to reach about 240 million, on average, for every combination to have been bought. This figure is nearly twice as many as the maximum sales ever achieved. Unless a syndicate deliberately organizes itself to buy large numbers of *different* tickets, avoiding the replication that arises when individuals make their own independent choices, some combinations will always remain unbought in any given draw.

To appreciate why total sales would need to be as large as indicated, think of what happens when every combination, except one, has already been bought. Each subsequent purchase has just one chance in 14 million of hitting this last combination, so, on average, it takes 14 million purchases to cover it. In a similar fashion, about 7 million purchases would have been needed for the last-but-one combination, and so on. These last few tickets take an awful lot of tracking down.

If you asked passers-by to tell you the month of their birth, or their star sign, how many would you expect to ask until all 12 answers were represented? If all 12 periods are the same length, and birth dates were uniformly distributed over the year, then the above formula with $N = 12$ would apply. The average number would be about 37. We have observed that any move away from uniformity tends to *increase* the average time to collect a full set. Months vary in length from 28 to 31 days, star signs vary a little less, and there is a summer birthday bias. Since we expect some variability about the average anyway, it seems most appropriate to work with round numbers, and suggest that we shall have to approach about 40 people to find one born under every star sign, or all 12 months represented in their birthdays.

As well as estimating this average number, it is useful to know how typical the average is. In Appendix III, we offer the notions of *variance* and *standard deviation* as ways to measure variability around an average, and note that we can often use the Normal distribution as a good approximation. At such times (and this is one of them), the actual value is

- about equally likely to be above average, or below it

- *within one* standard deviation of the average about two times in three

- *more than two* standard deviations away from the average only about one time in 20.

For example, if the average is 100 and the standard deviation is 10, then two-thirds of all values will be between 90 and 110, with one in six below 90, and one in six above 110. Only one value in 20 will be outside the interval from 80 to 120. But if the standard deviation were 30, the spread will be much wider. One-third of all values would fall outside the range from 70 to 130.

Table 7.2 Estimates of the total number of purchases required for a complete collection of all N objects.

N	12	24	48
Average	37	91	214
Standard deviation	14	29	60
2/3 of the time	23 to 51	62 to 120	154 to 274

Full details of the required computations can be found in textbooks; an outline is given in Appendix III. Assuming all N 'coupons' are equally likely with any purchase, and N is ten or more, the standard deviation of the total number of purchases required for a full set is a little less than

$$N\pi/\sqrt{6}$$

(another unexpected bow from that ubiquitous number, π). To a good approximation this standard deviation is just under $1.28N$. Table 7.2 shows these formulae in action.

Draw your own conclusions: quoting averages to any number of decimal places is plainly inappropriate, whatever the value of N. Actual values well above, or well below, the quoted averages should not be surprising.

But if the distribution of coupons is non-uniform, the calculation can change radically. The average number of purchases will be larger than given above, and will be dominated by the need to acquire the rarities. For example, suppose you seek to collect 12 different objects, 11 of which occur equally often, but one rare object is found only once in 100 purchases. On average, we shall need 100 purchases to get that rare object. But then, if it did take anything like 100 purchases to obtain the rarity, we would be desperately unlucky if we had not then *already* obtained all the more common objects. We saw that, if all 12 were equally frequent, we would need to make 37 purchases on average, and only one time in 40 would we expect to need more than 65.

This shows a nice paradox: the problem is more complicated when the objects have different frequencies, but estimating the average number of purchases can be easier. If one object is *much* rarer than the others, the average number to acquire the whole set is only marginally more than the average to get that one alone. Moreover, the variability of the number of purchases is also larger, so there is even less point in seeking an exact calculation of the average. With one very rare object, and any number of rather more common ones, replace the whole problem with the simpler question: how long to wait to get the one very rare object?

The secretary problem

You are a high-powered executive who makes instant decisions. You need a new secretary, and the applicants are queuing outside your office suite. In turn, each can be given a brief interview, at the end of which you will summarily decide whether to offer the job, and terminate the interview process, or to reject the applicant and proceed to the next candidate. You may not recall previously rejected applicants, and you must offer the post to someone. How should you reach your decision?

This problem is reminiscent of that of the prince (or princess) in fairy tales who sees a succession of suitors. You may have faced a similar question as you travel in a strange town, glancing at the succession of pubs to see which you will choose for your lunch break, or at the parade of motels enticing you with satellite TV or eat-all-you-like breakfasts. To stop, or to press on? How do you choose?

Suppose your aim is to maximize the chance you will select the best candidate (replace by suitor/pub/motel to taste). If there are 60 candidates, random choice would give you a one in 60 chance of finding the best—there must be a way to do better than that! If we have no reason to believe otherwise, it seems reasonable to assume the candidates are presented in a random order, unrelated to their qualities.

Try the general policy of interviewing a proportion of the candidates to give a feel for their overall suitability, and then selecting the first candidate who is better than all of these rejected. At worst, you will have to accept the last applicant, who has a one in sixty chance of being the top one anyway. I claim that the policy of rejecting the first half of the candidates, and then subsequently accepting the best so far, will pick the best candidate more than a quarter of the time.

To see this, consider where the *second best* candidate will be placed. By symmetry, s/he is equally likely to be in the first half or the second half of the queue of applicants. Half the time s/he will be in the first half, and get rejected. Suppose this happens. If the best candidate of all is also in the first half, you have missed her/him. But if s/he is in the second half, s/he *must* be accepted: s/he is the only candidate better than someone already rejected! Moreover, it is slightly more likely than not that s/he will be in the second half, because one of the places in the first half is spoken for. Overall, a little more than a quarter of the time, the second best candidate is in the first half, the best is in the second half, and the best is inevitably selected. There are other arrangements that also mean you would select the top applicant. This policy ensures your chance of picking the best is at least 25%.

We can do even better. We can increase the chance of picking the top candidate to about 37%. Our method is basically the same as just described, but we use fewer than half the candidates as a training sample. What you should do is to assess the quality of the first 37% (or so) of applicants, but reject them anyway. Then choose the first candidate, after this cohort, who is the best so far. The boxed argument explains why looking at about 37% of the applicants as a training sample is better than using rather more, or rather fewer.

May your pub meals be more palatable, and your motel stays more comfortable, from your use of this strategy.

How to choose the best

Suppose there are N candidates, and you want to explore what happens when you reject an initial number, and then choose the first candidate better than all of these. Think of N as being at least 6. Let Jo be the best candidate of all, and use the name Polly to mean the best candidate seen so far, before the present one.

Suppose we reject the first r candidates. Jo is in one of N positions, each with probability $1/N$. If she is in the first r places, we shall reject her. But if she is at position $r + 1$, we are certain to select her. If she is at $r + 2$, we cannot be sure, it depends on when we saw Polly. We shall choose Jo whenever Polly was in the initial r candidates—which has chance $r/(r + 1)$. Now suppose Jo is at $r + 3$; again, we select her when Polly was in the initial r, which now has chance $r/(r + 2)$. Continue in this fashion, placing Jo at $r + 4$, $r + 5$, ... , N. Adding up these disjoint chances, we select Jo with probability

$$\frac{1}{N} \times \left(1 + \frac{r}{r + 1} + \frac{r}{r + 2} + \frac{r}{r + 3} + ... + \frac{r}{N - 1} \right)$$

Call this $p(r)$—the chance we select Jo when we reject r initially. Our aim is to find that value of r for which this probability is as large as possible.

Replacing r by $r + 1$ throughout this formula tells us $p(r + 1)$, the chance we select Jo if we reject $r + 1$ to begin with. To see which of these chances is bigger, look at the difference $p(r) - p(r + 1)$. That difference collapses down to

$$\frac{1}{N} \times \left(1 - \frac{1}{r + 1} - \frac{1}{r + 2} - ... - \frac{1}{N - 1} \right)$$

When r is small, this will be negative, which means that $p(r)$ is *less than* $p(r + 1)$. For larger values, the difference becomes positive, so $p(r)$ then *exceeds* $p(r + 1)$. Thus, as we pass along the sequence $p(1), p(2), ... , p(N - 1)$, these chances first increase, reach some maximum, and then decrease. Where is that maximum? That choice gives us the best chance to select Jo.

The place where the maximum occurs is at the *largest* value of r such that

$$\frac{1}{r} + \frac{1}{r + 1} + \frac{1}{r + 2} + ... + \frac{1}{N - 1}$$

exceeds 1. Make this choice of r. The chance of selecting Jo is then

$$\frac{r}{N} \times \left(\frac{1}{r} + \frac{1}{r + 1} + \frac{1}{r + 2} + ... + \frac{1}{N - 1}\right)$$

which is plainly very close to r/N since the expression in brackets is very close to 1.

A bit of A level (double) maths tells you that r is about N/e, where e is the base of natural logarithms, so that r/N is about $1/e$. In percentage terms, that corresponds to the claims made about 37%.

CHAPTER 8

Let's play best of three

If two people, or two teams, compete for a prize, how many matches should they play? Once, the world snooker champion defended his title over a series of 143 frames; in modern times, up to 35 frames are used. The finals of the World Series at baseball, and the Stanley Cup at ice hockey, are played as best of seven matches, split four to three between the home venues of the contesting teams. Test cricket, in which drawn games occur frequently, is played over a series of from two to six matches. Chess world championships are seldom played under exactly the same rules each time, but 'best of 24' has been common recently. A single *game* at lawn tennis does not have a fixed number of points: to win, you must score at least four points, and you must also be at least two points ahead of your opponent. To win a *set* you must win six games, and be at least two games ahead (unless a tie-break operates); a *match* is usually either best of three sets, or best of five sets. In squash, played to traditional rules, there is a distinction between *winning* a point, and *scoring* a point; only the server can score a point, and whoever wins a point serves to the next point. The winner of a set is the first to score nine points, except for a twist near the end: if the score ever reaches 8–8, the non-server 'sets' the winning post either at nine points, or at ten points. Effectively, the non-server is allowed to decide 'let's play best of three', rather than 'next score wins'. Like lawn tennis, matches are played over three or five sets. Badminton has the same broad picture as squash, but with different winning posts, and different rules for 'setting'. What are the advantages and the drawbacks of the various formats? Can probability be useful in indicating good tactics?

The usual reason to have more than one contest is to reduce the element of chance, to give more confidence that the winner has demonstrated a real superiority. To seek to identify the better of two backgammon players through one game, or the better bridge team over just one hand, would be absurd. However, practical considerations will sometimes indicate that just one contest should be used to decide the winner. In professional boxing, there are sound medical reasons to demand a reasonable interval between bouts; the World Cup at soccer might become even more a test of endurance than of skill, if the single final were replaced by a series of matches. Statisticians seeking to give advice in this area should remember their humble place: the rules of games are designed to provide contests of reasonable length, and enjoyment for the participants and spectators. Games were not invented to be statistically efficient ways of deciding between two alternatives.

Contests should be seen as fair to both players, as far as is possible. Thus the advantage of White in chess is neutralized by allowing each side to hold the White pieces equally often, and so there will usually be an even number of games. The rules for winning a set in lawn tennis ensure that each player has the same opportunity to hold the advantage of the serve. No such fine tuning is needed in symmetrical games such as backgammon, soccer at neutral venues, arm wrestling, or tiddlywinks. Our aim in this chapter is to assess what difference the number of contests makes to either player's chance of victory.

Snooker-type games

The following model for successive frames in a snooker match will also be useful in other contests. Assume that:

- each frame is independent;
- the probability that either player wins a frame remains constant over a series.

Before you seek to apply the results we obtain, either to snooker itself or to other games, consider carefully whether these assumptions are reasonable; if they are not, our analysis should be distrusted. Take lawn tennis: at the low standard I play, it matters little who is serving, and these assumptions translate very reasonably to successive points in the whole match. But even at club level, let alone Wimbledon or Flushing Meadow, players win far more points when they are the server, so the assumptions above could apply at best only to points within the same game, and not to points over several games. Nor would these assumptions apply to a series of horse races, if the winner is handicapped in later races. In baseball or ice hockey, if there were an appreciable home advantage, this would not be a good model for the World Series or the Stanley Cup. However, provided each 'match' in a series is played under very much the same conditions, and there is no good reason to believe recent results affect the chances of winning the next match, this model for frames of snooker should give good guidance.

We want to see how the chance of winning a frame translates into the chance of winning a match, for matches of various lengths. Suppose that one player, Smith, has probability p of winning any frame, and the winner of the match is the first to win N frames. In snooker, no frame can be tied. Most interest will be where the two players are well matched, with p near to 1/2. So write $p = 1/2 + x$, and think of x as being small.

The match normally ends when either player has won the N frames required. However, it plainly makes no difference to the identity of the winner if we were to allow the game to continue to its maximum possible length—there are insufficient frames left for the loser to catch up. (When Joe Davis was world snooker champion, matches did indeed continue for their full length, even after the winner had been decided! Television today would balk at such indulgence.) This convention means that all games can be treated as having exactly the same length. So if the winner is the first to win five frames, we will imagine extending all matches to their maximum span of nine frames; if the contest target is N frames, all contests will last for $2N - 1$ frames.

This now sets the match up in the standard framework of the binomial distribution (Appendix II), because:

- there is a fixed number, $2N - 1$, of trials (frames);

- each trial is independent;

- Smith's chance of winning each trial is fixed at $p = 1/2 + x$;

- the quantity of interest is the total number of trials Smith wins—is it at least N?

Using Appendix II, we can write down the chance that Smith wins any particular number of frames. We then calculate his chance of winning the match by adding up the chances for all scores in which he wins at least N frames.

Playing best of three corresponds to being first to win $N = 2$ trials. Smith wins all three trials with probability p^3, and wins exactly two trials with probability $3p^2(1-p)$, so his chance of winning the contest is the sum of these two expressions. Re-writing the answer in terms of x, this chance is $\frac{1}{2} + \frac{3x}{2} - 2x^3$. When x is small, x^3 is tiny, so the first two terms give an excellent approximation to the final answer. Table 8.1 shows how the chance

Table 8.1 The probabilities of winning a best-of-three contest, for different chances of winning a single game.

P(Win one game)	0.52	0.55	0.6	0.67	0.75	0.85
P(Win match)	0.53	0.575	0.643	0.745	0.844	0.939

of winning a single game translates into the chance of winning a best-of-three contest. When p is just over one half, Smith's winning chances over best of three are only marginally more than in a single frame. Best of three should be regarded as helping to rescue a firm favourite who slips up at the first frame, rather than a serious attempt to separate two closely matched contestants.

Another way to show this same information is via the graph in Figure 8.1.

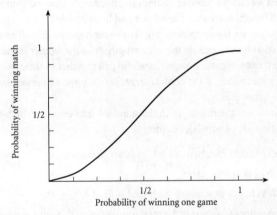

Fig. 8.1 The chance of winning a best-of-three contest, given the chance of winning one game.

Since many sporting contests are played as best of a series of either three, five, or seven 'sets', it is convenient to note how the chances of winning a *set* convert into the chances of winning a *match*. Assume successive sets are independent. The exact expressions are given in Appendix II, but an illustration when the chance of winning a set is 60% is representative. In this case, the respective chances for matches with three, five and seven sets are 65%, 68%, and 71%.

The same method can be applied to contests of other lengths, to give an answer for best of nine, best of whatever. These answers do not collapse down to neat or succinct expressions, and I will not give the details.

However, when x is small, so that the contest is close, there *is* a good approximation. Whatever fixed value N takes, we can write out Smith's winning chance in terms of the powers x, x^2, x^3, etc.. For large N, these expressions involve factorials, and we can use Stirling's formula to simplify the answer. When Smith's winning chance on a single frame is $\frac{1}{2} + x$, his chance over a first-to-N contest turns out to be

$$\frac{1}{2} + x\sqrt{\frac{4N}{\pi}} + \ldots, \qquad (*)$$

The first omitted term is a multiple of x^3, because *all the terms in x^2 cancel out*. This means that for small values of x, the value of (*) will be an excellent guide to the exact answer. (When $x = 0.1$, then x^3 is only 0.001.)

This formula gives a good idea of how the target to win magnifies any advantage a player has over a single frame. Note that it is the *square root* of N, the target to win, that appears. As this target increases, its square root also increases, but *much more slowly*. To double the winning advantage, the contest needs to be four times as long! For example, if Smith would win 55% of matches with a target of five frames (best of nine), the target would have to be 20 (best of 39) to increase his winning chance to 60%. From the other perspective, it should make very little difference if the contest is first to 20 rather than first to 15.

For close contests, so that x is small and we can use the formula (*) with confidence, notice how slowly Smith's winning chance changes as the matches get longer. Table 8.2 gives relevant figures. Take the case when Smith wins 51% of frames. Then $x = 0.01$, so in a 'best-of-19' match, where the winner is the first to win ten frames, his chance of winning the match has increased only to about 53.5%. But if he wins 60% of frames, so that $x = 0.1$, he wins about 85% of such matches. In a best-of-seven contest, such as the World Series, the entry for $N = 4$ and $x = 0.05$ shows that a team that wins 55% of games will win overall nearly 61% of the time. If it can push its single-game chance up to 60%, it would win the Series about 71% of the time. When Smith tends to win just over 50% of frames,

Table 8.2 The chance Smith wins one frame is $1/2 + x$ (x small); for contests of different lengths, Smith's winning chance is $1/2 +$ the entry in the table.

First to:	1	2	3	4	5	7	10	16
Extra winning chance	x	$1.5x$	$1.875x$	$2.1875x$	$2.461x$	$2.933x$	$3.524x$	$4.478x$

Table 8.3 The probabilities of winning a contest of first-to-*N*, for various values of *p*, the probability of winning a single frame.

Target *N*	5	10	15	20	25	30
$p = 0.52$	0.55	0.57	0.59	0.60	0.61	0.62
$p = 0.55$	0.62	0.67	0.71	0.74	0.76	0.78
$p = 0.60$	0.73	0.81	0.86	0.90	0.92	0.94

a very high target is needed before his overall winning chance noticeably exceeds 50%.

Be careful in the use of (*). For any fixed *x*, its value will exceed 1 when *N* is large enough, which would be ridiculous as no probability can be bigger than 1. Given any *x*, (*) can only be used for a limited range of *N*. You should be safe if you use it only when *N* is less than about $1/(10x^2)$.

Table 8.3 demonstrates how many frames might be required so that the better player has a substantial chance of winning the match. In the snooker World Championship, better players are seeded, and the first round is played first-to-10; so for a seeded player to have an 80% chance of winning, he should expect to win nearly 60% of frames, on average. The target increases in later rounds, being first-to-16 in the semi-final and first-to-18 in the final; at these stages, we can expect players to be very closely matched—e.g. that *p* be between 45% and 55%—so the slightly inferior player has a chance of more than one in four of causing an upset.

In any serious match, the number of frames is fixed beforehand. But, between friends, you can be more flexible. If the Emperor Nero had played snooker, he would surely have constantly re-negotiated the target to win during the play. 'Nero's rules' are: if you win the first frame, you declare yourself the winner of the contest; but if you lose it, you change the rules to 'best of three'; if you are still behind after three frames, you similarly adjust the length to five, seven, or more. In this dictatorial set-up, you are allowed to claim victory should you *ever get ahead*. Under these favourable circumstances, what is Nero's chance of victory?

Using the same model as above, there are three cases to look at.

- Your chance of winning a frame, *p*, exceeds 50%.
- The value of *p* is exactly 50%.
- *p* is less than 50%.

The most straightforward is the first one. Here, in the long run, you win more than half the contests, so it is certain that if you play long enough, you will get ahead sometime. The same conclusion also holds when your winning frequency is exactly one half. This is because, over a long period, your actual winning proportion will fluctuate near its eventual value of one half. Sometimes these random fluctuations take your proportion of victories under one-half, but sometimes the proportion exceeds one-half—and the first time this happens, you triumphantly claim victory.

The interesting case is the last one. In the long run, the proportion of frames you win settles down to p, so you must eventually fall behind and remain behind. However, you *might* get ahead in the early stages. We seek the chance that you ever get ahead, so that you can declare the contest over. The answer turns out to be beautifully simple:

when p is less than one-half, your winning chance is $p/(1-p)$

Suppose, for example, that you win one frame in three. Then $p = 1/3$, so that $1 - p = 2/3$, and the chance you ever get ahead is 1/2. If p is only just under 50%, you have an excellent chance of getting ahead sometime; when $p = 45\%$, your chance is 45/55, about 89%. But if p is as small as 10%, your winning chance is only 10/90, about 11%. The box shows how this neat answer arises, and also formally confirms that you are certain of victory when p is at least one-half.

The chance you ever get into the lead

Suppose X denotes the chance you ever take the lead in a series of contests, as the scores accumulate. Your chance of winning a single contest is p. After the first frame, then either

- with probability p you have won it, so have won the contest immediately, or
- with probability $(1 - p)$ you have lost it, so you are now one frame behind.

Consider how you might win in the second case. Since you are now one behind, you must:

(1) sometime make up that deficit to bring you level
(2) then subsequently go on to take the lead sometime.

But making up a deficit of one to bring you level requires exactly the same performance as getting ahead by one from a level start, so the chance that (1) occurs is also X. Moreover, once you have drawn level, your chance of a subsequent lead is X; so the chance of both (1) and (2) is X^2.

Collecting these together,

$$X = p + (1-p)\,X^2.$$

Using standard methods to solve quadratic equations, the two possible solutions are $X = 1$ and $X = p/(1 - p)$. Note that X is a probability, so any answer with X greater than 1 is impossible. If p exceeds $1/2$, that second answer is greater than 1, so the answer we seek is the other one, $X = 1$. If $p = 1/2$, both possible solutions are equal to 1, so $X = 1$ again. This confirms that if p is at least $1/2$, then $X = 1$ and you are indeed certain to take the lead sometime.

When p is less than $1/2$, both solutions *could* make sense. But if you are prepared to accept (as is the case) that you cannot be absolutely certain of getting ahead, only the second solution, $X = p/(1 - p)$, is tenable.

Squash-type games

The traditional scoring method in squash allows a player to score a point only if she is the server. We will examine the development of games that use this rule, while acknowledging that many professional tournaments are now played under the rule that the winner of a point scores, irrespective of who is server. In the traditional method, the first task is to see how the probability p of *winning* a point translates into the probability of *scoring* a point. For simplicity, we assume p is the same, whoever serves. Write

S = the probability you score the next point, if you are now the Server

R = the probability you score the next point, if you are now the Receiver.

The boxed analysis leads to expressions for both S and R in terms of p. So if we know the chance of winning any point, we can convert it to give the chances of scoring a point, whether you are currently serving or receiving.

Figure 8.2 shows how S and R are related to p. Notice how much larger S is than R over the middle range of values of p. This helps explain why it is not unusual to find set scores as extreme as 9–0, 9–1, or 9–2, even between well-matched players. Even when $p = 1/2$, then $S = 2/3$, so the server has a good chance of scoring several points in succession.

If the score reaches 8–8, and you are the Receiver, should you exercise your right to have the game played as first to ten, or leave it as first to nine?

Converting winning points to scoring points

If you are now the Receiver, the only way you can *score* the next point is first to win the next point, and then score from the position of being Server. So

$$R = pS$$

If you are now Server, you can score the next point in two ways: either you win the next point (probability p), or you lose that point and become Receiver (probability $1 - p$), but then score the next point from being Receiver. So

$$S = p + (1 - p)R.$$

These two equations lead to the answers

$$S = p/(1 - p + p^2), R = p^2/(1 - p + p^2).$$

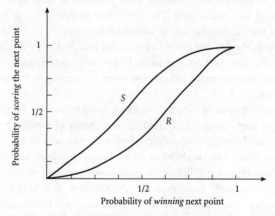

Fig. 8.2 The chances S and R of scoring the next point, according to whether you are currently Server or Receiver, when the chance of winning any point is p.

To decide, we have to work out your respective winning chances. If you leave it as first to nine, you win only when you score the next point. As you are currently Receiver, your chance of doing so is R, whose value is shown in the box. Suppose you exercise your right to set to ten: in terms of the next points scored, there are three distinct sequences that lead to victory, namely:

- Win, Win
- Win, Lose, Win
- Lose, Win, Win.

The first has chance $R \times S$ (you must win a point as Receiver, then as Server). The second has chance $R \times (1 - S) \times R$ (win as Receiver, lose as Server, then win as Receiver again). The last has chance $(1 - R) \times R \times S$ (by a similar argument). Add these three values up to find the chance of winning.

Which alternative gives the greater winning chance? Suppressing the details, the answer reduces to a very neat form: it is better to set to ten whenever the value of S exceeds one-half. The box shows how S is related to p, the chance of winning a point, so we can find what this means in terms of p. In round figures,

opt for setting to ten whenever p exceeds (about) 38%.

In practice, this will happen nearly every time: you only have the opportunity to exercise this choice if the score has reached 8–8, which indicates the two players are well matched. Thus p should be near 50%. Almost certainly, it will exceed 38%. The most likely exception would be when you, but not your opponent, are close to exhaustion, or have just picked up an injury. In this case, p will be small, and your best hope of winning is to make the game as short as possible. If you assess your current chance of winning a point as under 38%, set the target to nine; the rest of the time, make it ten.

Just as in frames of snooker, we can calculate how the frequency of winning a single point is converted into the chance of winning a set. We have just seen that the probabilities of the two sequences Win, Lose, Win and Lose, Win, Win are different, even though they both lead to a 2–1 score. Contemplate with horror the corresponding task of enumerating all the sequences, with their respective probabilities, that lead to squash scores such as 9–6, or 10–8, and recoil! There must be a better approach.

The basic model assumes that your chance of winning a point is the same whether you are Server or Receiver. Against a well-matched opponent, you fill both roles about equally often. In order to *score* nine or ten points, you have to win that number as Server, but you might also expect to win about as many when you are Receiver. This rough calculation suggests that your target is to win about 18–20 points altogether, in order to win the set. If you are slightly better than your opponent, you will expect to win more points, so a more realistic target might be a little less, say 16–18 points, so as to reach the winning score of nine. (When players play the rule that each point scores, irrespective of server, the target score is normally 15.)

This rough argument indicates that for p in the range from 40% to 60%, squash can be analysed in the same way as snooker, with the target-to-win N being about 16–18. The use of (*) on p. 149, or Tables 8.2 and 8.3, would then lead to estimates of the chance of winning a set. Computer simulation backs this up, with N replaced by about 16–18 when p is between 45% and 55%. There is no call to seek more precision: we know the model has imperfections, and we are hoping to find broad indications, not to quote over-precise answers. Moreover, the way this chance depends on N is mainly through its *square root*; $\sqrt{16}$ and $\sqrt{18}$ are very similar. Having found the chance of winning a single set, this can be converted to the chance of winning a best-of-three or best-of-five match by using Table 8.2, or the formulae in Appendix II. For example, if $p = 51\%$, the chance of winning a set is about 55%; Table 8.2 with $p = 55\%$ converts this to a winning frequency of about 57.5% in matches over three sets, and to about 59% in matches over five sets.

Anyone for tennis?

The scoring system in lawn tennis is beautifully designed to ensure that crucial points occur frequently. No matter how long the match has gone on, and how far behind she is, the player losing can still win. ('It ain't over till the fat lady sings.') The game also has the anomaly that even a set, let alone a match, can be won by the player who wins fewer points! One way is when you win six games, which all went to deuce, while your opponent whitewashed you in the other four games: you are victorious, she scored four more points.

Recall that a single *game* goes to the first player to win four points, with a lead of two points. It seems reasonable to suggest that, within a single game where the same player serves all the time, the snooker set-up applies, with 'frame' replaced by 'point'. Write p as the probability Smith wins any point, and for convenience write $1 - p = q$, so that q is the probability Smith loses a point. If $p = 1/2$, total symmetry shows that Smith's chance of winning the game is also 1/2, so we can concentrate on the case when p is not 1/2. The boxed argument shows how to convert the chance p of winning a point to the chance G of winning a game.

Figure 8.3 (next page) shows how G increases as p increases. What is striking about this graph is how quickly G increases when p is only just above 1/2. Indeed, by the time p has reached 70%, then G is up to 90%. This points to a law of rapidly diminishing returns: if you win 70% of

From a point to a game

First, we will find Smith's chance of eventually winning the game, if the score is now deuce. Denote this chance by D, and consider the next two points. With probability p^2 she wins them both, and so wins the game; with probability q^2 she loses both, and has lost the game; otherwise, with probability $2pq$, she is back where she started at deuce. This gives the equation

$$D = p^2 + 2pqD,$$

which leads to $D = p^2/(p^2 + q^2)$ as the chance of winning from deuce.

Smith's chance of winning a game, G say, is found by listing the distinct scores she can win by, and adding up their probabilities. These are

- win four points, probability p^4
- win three of the first four points, then the fifth—probability $4\,p^4 q$
- win three of the first five, then the sixth—probability $10\,p^4 q^2$
- win three of the first six, then win from deuce—probability $20p^3q^3D$.

Adding them up, and writing the answer in a neat format, we find

$$G = \frac{p^4 - 16\,p^4 q^4}{p^4 - q^4}.$$

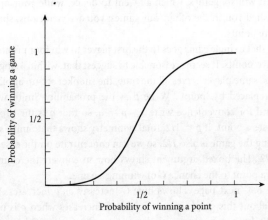

Fig. 8.3 The relationship between the chance of winning a game at tennis (G) and the chance p of winning any point.

points on your own serve, you will already win 90% of these games, so working to improve your serve even further will make little difference to the number of games you win—it is already very high. Data from the Wimbledon Championships in 1997 show a range of plausible figures, consistent with this formula. On service, Henman won 75% of points and 22 service games out of 23; his opponent, Krajicek, won 73% of service points and 21 service games out of 23. In the Ladies' Final, Hingis won 55% of service points, and nine of thirteen games while Novotna won 54% of service points, and eight of thirteen games.

To examine more closely what happens when p is near $1/2$, write $p = 1/2 + x$, and again think of x as being small. The boxed expression for G can be written as

$$G = \frac{1}{2} + \frac{5x}{2} - \ldots;$$

once again, the first omitted term is a multiple of x^3. This expression has one very useful consequence when p is close to $1/2$: the multiplier '5/2' indicates that if Smith makes a *tiny* improvement in her chance of winning a single point, this converts into an improvement of nearly 5/2 times that amount, for her chance of winning a game. If you increase your chance on a single point from 50% to 52%, you increase the proportion of games you win from 50% to 55%.

Above, we looked at the relation between the chances of winning a point and of winning a game. What about the relation between winning a game and winning a set? Service alternates, and the first to win six games, and be two ahead, is the winner. We should assign different values to the chances of winning a single game according to who is server. Suppose Jones wins a game when he serves with probability s, and when he receives with probability r. The data above from 1997 Wimbledon show that values of s in excess of 90%, and r below 10% can be expected.

Before tie-breaks were introduced, matches could reach a great length through set scores such as 15–13 or 24–22. Some tournaments insist that a tie-break is not applied in the fifth set of five-set matches, or the third set in best of three. For a set with no tie-break rule, Jones's winning chance is found using arguments similar to those that gave our earlier formulae. The answers are inevitably more cumbersome, because there are now two probabilities, r and s, instead of the single probability p for every point. It is not worth stating the rather complex expression that results, but we can use it to paint a broad picture.

The key quantity turns out to be the *average* of the two probabilities s and r. Write this average as $P = (s + r)/2$. If P is exactly one-half, his chance of winning the set is also one-half, whatever the actual values of s and r. To have a better than 50–50 chance of winning a set with no tie-break in use, we require P to exceed one-half. This simple criterion will confirm the instinct of a good tennis coach: it is more profitable to eliminate weaknesses than to hone your strengths yet further. Because, suppose your serve is so good that $s = 90\%$; no matter how much better your serve gets, s cannot exceed 100%. But if r is down at 10%, the potential to increase it is much larger. It seems more realistic to hope to increase r from 10% to 15% than to increase s from 90% to 95%—but these two changes have very similar effects on your chance of winning a set.

In a tie-break, one player serves to the first point, then the players take turns to have two successive serves. The winner is the first to win seven points, and to be two points ahead. To find the winning chances here needs identical calculations to those already used to win a set, except that the initial target is now seven instead of six. Using s and r to indicate the chance that Jones wins a *point* as server and receiver, the key quantity is still $(s + r)/2$. Data from the US Open in 1997 confirm how closely matched the top players are. In one semi-final, Rafter beat Chang by the convincing score of 6–3, 6–3, 6–4, and yet the points score was 97–78—the loser took 45% of the points. In the other semi-final, Rusedski beat Bjorkman 6–1, 3–6, 3–6, 6–3, 7–5; plainly a close match, shown even more starkly by the points total, which Rusedski won 128–127! Tables 8.4 and 8.5 give illustrations of the chances of winning sets and tie-breaks when s and r take values typical of top class tennis.

Legislators want games to be seen as fair. After an odd number of points in a tennis tie-break, one player has had one more serve than the other. But the arrangement that players have two serves in succession, after the first point, means that they take it in turns to be the one who has had the extra

Table 8.4 Let s and r be the probabilities Smith wins a point as server or receiver respectively. The entries are Smith's corresponding chances of winning a tie-break.

$s=$ / $r=$	0.5	0.4	0.3	0.2
0.5	0.500	0.345	0.206	0.098
0.6	0.655	0.500	0.337	0.185
0.7	0.794	0.663	0.500	0.317
0.8	0.902	0.815	0.683	0.500

Table 8.5 Here s and r are Smith's chances of winning a *game* as server or receiver. The entries are the chances of winning a set at tennis, with no tie-break rule.

$s=$ / $r=$	0.5	0.4	0.3	0.2	0.1
0.5	0.500	0.354	0.221	0.113	0.038
0.6	0.646	0.500	0.347	0.200	0.079
0.7	0.779	0.653	0.500	0.327	0.150
0.8	0.887	0.800	0.673	0.500	0.275
0.9	0.962	0.921	0.850	0.725	0.500

serve. In contrast, service alternates every game in a set, so the one who was receiver in the opening game will never have held the service advantage more often.

Our mathematical model with its probabilities s and r for winning as server or receiver leads to the same chance of winning a set (or tie-break) *whichever method of alternating serve is used*. All that matters is that after an odd number of points (or games), one player has had the advantage just once more often than the other, and to win you must be two points (or games) ahead of your opponent. But tennis buffs speak of the 'advantage' in a set without a tie-break of being with the one who served to the first game. The idea is that if you serve first in the set, and lose your serve any time after the score has reached 4–4, at least you have a chance to break back. But if games go as with service for nine games, and you are serving second in the set, any time you lose your serve you have then lost the set. There is no chance to come back. If there is such an advantage, here is a simple way to eliminate it:

Apply the rule that alternates serves in a tie-break also to games within a set.

Summary

One contest may be seen as too brief to be an acceptable way to pick a winner. Where it is practicable, a series of contests is often arranged, with the first to win N being declared the winner. We have sought to relate the chance of winning an individual contest to the chance of winning a match consisting of such a series. It is a vital part of our model that the chance of winning the next contest is unrelated to the current score. We assume players do not become so demoralized when they are behind, or so over-confident when ahead, that their play is affected.

If players are closely matched, then a match of best of three, or best of five, makes very little difference to the chance that the (slightly) better player wins. The main effect of such a short series is to allow a stronger player a second chance, should she lose the first game. When one player wins 60% or more of games, they win at least 80% of contests when the winning target is ten or more. But when one player is only marginally superior, it will take a very long contest to increase her winning chance to that level. The rate at which supremacy in a single game is converted to supremacy over a long match rises only as the *square root* of the target to reach.

The analysis is modified to deal with contests such as traditional scoring in squash or badminton, where only the server can score a point, and to lawn tennis, where the chance of winning a point changes greatly according to who is serving.

CHAPTER 9

TV games

Game shows are popular on television world-wide. Some quiz shows rest entirely on the accumulated and arcane knowledge of the contestants while, at the other extreme, some games are completely a matter of chance, with no opportunity to combine luck with skill. I will describe four games where ideas from probability can be used either to improve your winning chances, or to gain a better understanding of the game.

Find the lady

In this well-known game, there are three identical boxes, A, B, and C. One of them contains a substantial prize, the other two are worthless. The contestant's task is to guess the correct box. The host, Hugh, knows which box is the valuable one. After some preliminary skirmishing, the contestant, Mary, selects one box, B, say.

Whatever box Mary selects, at least one of the remaining boxes is empty. Hugh opens *one* empty box, and invites Mary to change her mind. Almost inevitably, she declines to do so. She knows that Hugh can always open an empty box, and so reasons that his doing so makes no difference. She is wrong. Provided—as seems to be the case—Hugh will always seek to tempt her by showing an empty box, she *doubles her winning chances if she switches*.

Many people who should know better continue to disbelieve this statement. Marilyn vos Savant, said to have the highest IQ in the world, gave a correct explanation to USA readers of the 'Ask Marilyn' column in *Parade* magazine. Her postbag bulged with excited letters, some from professors in university statistics departments, asserting that she was wrong. But her analysis was not only valid, it is disarmingly simple.

The easy part is to consider what happens if Mary *sticks* to her first choice. Since she was selecting at random amongst three boxes, her chance of being right is one in three.

Two times in three, Mary's initial selection is empty, but she does not know this. When this happens, Hugh has a problem. Of the two boxes left, one contains the prize, the other does not. It would be silly of him to show Mary the prize by opening the correct box. He *must* open the other empty box, so that when Mary switches, she is inevitably guided to the prize. Whenever her initial choice is wrong—two times in three—her final choice is right. She doubles her chances from one-third to two-thirds.

Tempting fate, I persuaded Topical Television to try this out in Autumn 1996. They found a group of 53 children from Crofton Hammond Junior School, Stubbington, willing to be guinea pigs. Each child played both the roles of sticker and switcher once, and the most numerate girl in the class tallied the scores. According to the analysis above, the average numbers of stickers and switchers who guessed correctly would be 17.67 and 35.33 respectively. What a relief to find that 19 when sticking, and 36 when switching, located the prize! Thank you, children, for your convincing compliance with statistical theory.

It is a vital part of the analysis that Hugh will always show an empty box, irrespective of Mary's choice. If he opened an empty box only when Mary had chosen correctly, it would be disastrous for Mary to switch. But the conventions of this as a TV game seem to demand that Hugh acts as described.

Just as some punters may continue to believe that the collection {1 2 3 4 5 6} must be selected less often than average in the National Lottery, so this description will not convince everyone of the benefits of switching. Should you not be convinced, consider the same game, but with 100 boxes, and just one prize. You make your guess, and Hugh, who knows where the prize is, will open 98 boxes and show them all to be empty. Whichever box you selected, he will always do this. He then invites you to switch.

If you refuse to switch, you win the prize only when your first guess was lucky, one time in one hundred. The other 99 times, when you failed to pick the winning box, Hugh's 99 boxes consist of the winning box, along with 98 empty ones. He must open these 98 losing boxes, and leave the one prize-winning box for you—if you switch. The policy of switching is equivalent to betting that the winning box is among the collection of 99 boxes—which happens 99 times in 100.

If you will agree it is better to switch in this game, why not also with three boxes?

As a complete aside, but where the same sort of probability argument with partial information is involved, try the following problem. A family is known to have two children. We want to work out the probability that both are Boys, or both are Girls, or there is one Girl and one Boy. It is assumed that Boys and Girls arise equally often, on average, and successive births are independent (our family is *not* a pair of identical twins). There are four equally likely possibilities, as the family may have arisen as BB, BG, GB, or GG. So the correct answer is that the respective chances of two Boys, two Girls or one of each are 1/4, 1/4 and 1/2.

How should we deal with the information 'The elder is a Boy'? In this case, the younger is equally likely to be B or G, so we can conclude that two Boys, or one of each sex, each have probability 50%. We get the same answer if we are told 'The younger is a Boy'. So what happens if we are told 'The family contains a Boy'?

It is very tempting to argue: 'If one of them is a Boy, then either the elder is a Boy, or the younger is a Boy. In either case, we saw that two Boys, or one of each sex, are equally likely. So the same applies here—there is a 50% chance of two Boys, and a 50% chance of a Boy and a Girl.' That argument is false; the chance of two Boys is 1/3, whereas the chance of one of each sex is 2/3. A valid argument notes that the statement 'The family contains a Boy' excludes only GG from the four equally likely possibilities, {BB, BG, GB, GG}. With that excluded, the three remaining families,{BB, BG, GB}, are equally likely, and this leads to the answer given.

The similarity with the 'find the lady' problem is through the way the additional information changes the set-up. In 'find the lady', when Mary chose the winning box initially, Hugh could open either box; but when Mary chose an empty box, Hugh was left with only one possibility. In our family problem, the information 'The elder is a Boy' excluded both GB and GG; and to be told 'The younger is a Boy' excludes both BG and GG. However, 'The family contains a Boy' excludes only GG.

Wipeout

This TV game has three players and three stages. In stage 1, of which more later, one player is eliminated. Stage 2 is a '*Wipeout* auction' that eliminates a second player. At the last stage, the remaining player has a chance to win a substantial prize.

The details of the stages are different, but the main idea is the same. A grid of objects is presented, and the contestants are told *how many* of them possess a specified property. Their aim is to identify the objects having that property, using knowledge and intelligent guesswork. Imagine that Julie and Natasha are to contest a *Wipeout* auction, faced with the 12 names displayed.

David Copperfield	Ebenezer Marley	Obadiah Slope	Mr Pickwick
Mistress Quickly	Magwitch	Bob Cratchit	Edwin Drood
Heathcliff	Wackford Squeers	Nicholas Nickleby	Fagin

The contestants are told (as you have no doubt noticed) that exactly eight of these are genuine names in novels by Dickens, and four are not. First, they conduct an auction to see who wins the right to attempt to identify a certain number of correct answers, between one and eight. This auction, with the players' private thoughts in parentheses, might proceed as follows.

Julie: (Help, I don't know Dickens at all. I saw David Copperfield on TV— wasn't Magwitch in that as well? Heathcliff isn't Dickens. I'm not sure about any of the others. They'll all laugh if I say two. I'd better sound confident.) 'Four.'

Natasha: (I'm sure I recognize four, and Mistress Quickly is Shakespeare. I bet she could name four, and Obadiah Slope sounds a Dickens name.) 'Five.'

Julie: (Phew! Well, I won't get six. Fingers crossed.) 'Go on, name five.'

Natasha: 'David Copperfield.'   Audience cheers.
         'Bob Cratchit.'   More cheers.
         'Fagin.'   Cheers and applause.
         'Ebenezer Marley.'   Cheers and groans.

Host: 'Ebenezer Scrooge, yes. Jacob Marley, yes. But not Ebenezer Marley.'

At this stage, Natasha having failed to meet her bid, it is Julie's turn. She wins if she can identify just one further Dickens character. If she erred, Natasha would have a second chance, this time to give just one correct name, and so on. As soon as either gives one correct answer, the game ends.

Julie: 'Magwitch.'   Loud audience cheers.

Julie has won, even though Natasha knew more correct answers. Julie judged her opponent's psychology well, and pushed her too high.

How can mathematics help? Not a lot, but it can point you in the right direction. Your aim is either to win the auction with a low bid, or force your opponent too high. The lower your successful bid is, the higher your chance of winning the game. What would be the best tactics against an equally knowledgeable opponent? *You should make the highest bid B for which your overall chance of winning is at least 50%.*

To see why this is so, look at the consequences of all possible strategies. If you make this bid of *B*, and are allowed to play, you win at least 50% of the time; and if your opponent overcalls with a higher bid, you will then let her play—her chance is under 50%, so your chance exceeds 50%. But suppose you bid less than *B*, and she counters with a bid of *B*. She has now put herself in exactly the previous position that you were in when you bid *B*, so whatever you now do, your chance is at most 50%. Finally, if you bid more than *B*, she should let you play, and you will win under 50% of the time.

Read that last paragraph again. That sort of reasoning arises very frequently in contests between two people: to find out what your own best move is, put yourself into your opponent's shoes, and see how she would react to your possible plays. When we looked at games with few choices, we usually found our best choice in this way.

How should you assess your winning chance, so that you can identify what to bid? You will be sure that some objects definitely do possess the specified property, and that some others certainly do not. For the rest, you guess in varying degrees. Take the example when you are certain that you recognize *four* that possess the property, and *one* that does not. This leaves seven objects where you are uncertain, four of them do have the property, and three do not. To keep things simple, suppose that when you have to guess, you do so completely at random; you do not think that some objects are a bit more likely than others.

In this example, if you win the auction with a bid of four or fewer, you are sure to succeed. So suppose you won the auction with a bid of five. You will reel off the four you know, and then begin to guess. You guess right, and win immediately, four times in seven. Even in the other three times that your guess is wrong, you still have a chance. You have eliminated four correct objects and one incorrect one. You have left seven objects, of which four are right and three are wrong. You are sure you can identify one of the wrong objects. It is now the turn of Tina, your opponent.

If she has *exactly the same knowledge* as you, she will also be able to eliminate that same wrong answer, and so she will now guess right four times in six. The other two times in six, you get your second chance. This second chance comes round with probability $(3/7) \times (2/6)$. Of the six objects now left, four are right and two are wrong. Since you can eliminate one wrong choice, you will win from here four times in five. In total, this second chance evaluates to

$$(3/7) \times (2/6) \times (4/5) = 4/35.$$

You will not get a third chance, but your total chance of victory if your bid of five wins the auction is $4/7 + 4/35 = 24/35$.

Dare you bid six? Here there are three sequences leading to victory. You could name six correctly; you might, as above, make a mistake on your fifth, but still win; or you might get five right, the sixth wrong, but still win. Suppressing the details, and making the same assumptions, a similar calculation leads to a total chance of $17/35$, which is just less than one-half. So the right bid is five.

What should Tina do when she hears your call of five? We assumed her knowledge is exactly the same as yours. She is just as good as you are at these calculations, and she knows that if she lets you win this auction with your bid of five, you win $24/35$ of the time. That means her winning chance would be only $1 - 24/35 = 11/35$. On the other hand, if she overcalls you and bids six, she increases her chance to $17/35$. So if both of you play optimally, play and thoughts might run along the following path, using the Dickens frame shown above.

You: (I know Copperfield, Pickwick, Fagin and Nickleby. Heathcliff is not Dickens. I'm totally ignorant about the others. She won't let me get away with four, because she probably knows these names too. If I bid six, I'll be guessing twice, and my total chance of success is under a half. If I sound a bit unsure, she might let me have a go.) 'I'll try for five.'

Tina: (I can name four, and Heathcliff is Brontë. If she knows the same, she stands a good chance of getting five. I wanted to bid five, but she got there first. If I bid six, I'm unlikely to get it, but I should make the bid anyway, as that increases my chances. If I sound confident, perhaps she'll go for seven—then I'm in clover.) 'Six.'

You: (Fine. I would be an idiot to bid seven, I'll let her stew.) 'Go on, name six.'

This playlet is based on a number of assumptions, and precise answers such as '24/35' should be used as guidance only. If this approach has any use at all, you ought to reach similar conclusions if you used different assumptions. For example, perhaps both of you could name four correct characters, and identify one that is incorrect, but not exactly the same names. Or you may not be guessing entirely at random: maybe you are 80% sure that Squeers is a Dickens name, and 70% sure that 'Ebenezer Marley' is incorrect. So what you must do is test the *robustness* of your conclusions—do they change, if your assumptions change a little?

Suppose both of you know four Dickens names, and one that is certainly not Dickens, but your specific knowledge is different. Perhaps both of you know Fagin and Cratchit, she knows Copperfield and Drood, whereas you know Pickwick and Squeers. If Tina lets you win the auction with a bid of five, you will only succeed when Tina never has a chance to speak. You reel off the four you know, and then guess. The 4/7 of the time that guess is right, you win, but otherwise you lose, as Tina has a banker available. Bidding five keeps your chance above one-half. But should you be attempting to name six, you need two lucky guesses, which has probability (4/7) × (3/6) = 2/7, which is less than one-half. Once again, a bid of five is indicated.

Making parallel calculations when our knowledge is somewhere intermediate between complete certainty and total guesswork turns out to be quite complicated. When we embark on random guessing, and the truth about the guess we have just made is revealed, we are no better in the future—we are still flailing around at random. But suppose there are five objects left, three correct and two incorrect, about which we have varying degrees of uncertainty: perhaps we think object A is 90% certain to be correct, while B is 75% certain, and so on. Because there are known to be three correct objects, it makes sense that the probabilities we attach to the five objects should add up to 300%. We guess A, and find it is indeed correct; there are now only two correct objects out of four, so the new probabilities for the four remaining objects should add up to 200%. But there is no automatic way to adjust the previous assessment of 75% for B, and the probabilities given to the other three objects. Rather than offer some artificial formula, I will note this complication, and restrict the description to how an auction should proceed under the circumstances of our playlet, i.e.:

(1) Both players have *exactly* the same knowledge.

(2) That knowledge is either total certainty, or complete guessing.

(3) Each player knows x correct answers, and y incorrect answers.

(4) Both players play optimally.

These conditions are designed more for computational convenience than for realism, so giving a detailed prescription of the 'best tactics' for both players is not appropriate. However, working out the best tactics in our model is quite diverting! The summary below is a broad picture, slightly inaccurate in some places, where two alternatives have very similar results. But it is not misleading.

It is just as useful to be able to recognize with complete certainty four that are incorrect, as it is to recognize eight that are correct. So the cases when either $x = 8$ or $y = 4$ are obvious; bid eight and win. When you know seven correct answers, bid eight irrespective of the value of y. When $x = 6$, the first bidder should always bid seven, which the second bidder should pass, *unless* $y = 2$, when she does best to overcall with eight. When $x = 5$, you should bid six for low y, but bid seven for higher y; as responder, usually pass. When $x = 4$, the game should always be played to name six: for low y, bid five, and expect to be overcalled, and for higher y, bid six yourself. In the case where $x = 3$ and $y = 3$, bid six; be less ambitious for lower y. Would you ever be in a game where $x = 2$? It ought to be played to name either four or five. For $x = 1$, (surely not?) bid three for low y and five when $y = 3$. Finally, suppose $x = 0$; your powers of bluff will determine your fate—bid one fewer than when $x = 1$.

Do not pore over the detail: the assumptions behind these recommendations are specific, and in any real game, there will be significant deviation from them. But the pattern overall *is* worth discerning: the case when you are very sure you know all eight correct answers, or all four incorrect ones, plays itself. Otherwise, tend to bid one or two more than you are pretty sure about, especially when you can eliminate several plainly wrong answers. If you make a very low initial bid, you put your opponent in a position to dictate matters; do not bid only four if you intend to bid six should she overcall with five. Certainly overcall if you think your chance of winning with that higher bid is at least 50%; if you think she has a substantial chance of making her bid—at least 60% say—it is probably best to overcall, even when your own chance is a little below 50%. Do not forget the element of bluff. The best mathematicians in the world lose at poker to good poker players who think π is for lunch.

Earlier, I promised to analyse stage 1 of this TV game, where one of three players is eliminated. The mechanics here are that there are 16 objects, of which 11 belong to some specific family. The first player seeks to identify those that belong to the family; so long as he is correct, he accumulates increasing sums of money, and may choose to continue, or to pass to the next player. But whenever you give an incorrect answer, *all* your accumulated money is lost (hence the name, *Wipeout*), and play passes to the next player.

A round ends either when all five incorrect answers have been selected, or when all 11 correct ones are identified. The person eliminated is the one with least standing to his credit at the end of several rounds; in the case of a tie, the person with more correct answers over the whole game goes through.

It is quite common to find at least two players, and sometimes all three, have zero to their credit at the end. Wipeouts occur frequently. Thus, especially in the early rounds, it looks good tactics to attempt to answer a large number of questions, even at the risk of losing all your immediate gains, as you would be in an excellent position to profit from this tie-break rule. How much you could profit will be examined later. Your sole object is to survive the elimination series—any money won here is insignificant in the context of later prizes to the survivors.

An early round might be played as follows. Of the 16 objects in the display 11 are villages south and east of Huddersfield in West Yorkshire, five are from foreign parts. For those not familiar with these names, the misfits are denoted by *.

Skelmanthorpe	Scissett	Shelley	Cumberworth
Dykelands*	Kitchenroyd	Plush*	Brockholes
Farnley Tyas	Birds Edge	Knipton*	Shepley
Gilmorton*	Skirwith*	Lepton	Midgley

Alfred: (I'm not sure, I'll just keep guessing.) 'Brockholes.' 'Correct, £10.'
   (Continue.) 'Farnley Tyas.'   'Correct; and £20 makes £30 altogether.'
   'Skirwith.'   'Wrong, that's in Cumberland. You're wiped out.'
Betty: 'Skelmanthorpe.'   'Correct. Worth £30.'
   'Cumberworth.'   'Correct; and £40 makes £70.'
   'I think I'll pass.'   'Up to you: over to Chips.'
Chips: 'Gilmorton'   'Wrong, it's in Leicestershire. Back to Alfred.'
Alfred: (Why is Betty passing? Never mind.) 'Scissett.'   'Correct, £50'
   'Midgley.'   'Sixth correct answer gives you another £60.'
   'Lepton.'   'And £70 makes £180'

But now move ahead. Suppose several rounds have been played, and Alfred holds a clear lead in any tie-break over Betty, who is correspondingly well ahead of Chips. Alfred holds £40, Betty has zero, and Chips has £70. This is the final round of the elimination stage, and whoever stands last when it is over must leave the stage. The seven correct answers, and the two incorrect ones noted above have been offered, and all three players are guessing at random, any actual knowledge being spent. Skill is exercised by your choice of play or pass when your first guess is correct. Chips has just passed, and play continues:

Alfred: (Seven answers left, four are right and three are wrong. The odds are in my favour.) 'Dykelands.'   'Wipeout, sorry. That's in Scotland!'
Betty: (My chances are four in six.) 'Plush.'   'Wipeout, sorry. Dorset.'
Chips: (Great chance here.) 'Shelley.'   'Correct. You win £80.'

Alfred and Betty have zero, Chips holds £150, there are three correct and one incorrect answers left. The correct answers are worth £90, £100 and £110. *If all are guessing, how should the game continue, and what are their respective chances of being eliminated? Should Chips play or pass?*

If Chips can rely on Alfred to be on the ball, he should pass. The reasoning goes as follows. Since Alfred leads Betty in the tie-break, Alfred will be quite happy to guess wrongly, as the game will then end with Betty eliminated. Notice that if Alfred's first guess is correct, he should guess again—he must not pass because if he did, both Betty and Chips might make correct guesses and eliminate him. It is safe for Alfred to pass after two correct guesses, as he then has £190 and Betty will miss out, whether or not she guesses correctly. So optimal selfish play is for Chips to pass, and Alfred to guess up to twice. Betty will lose out.

But if Chips passes and Alfred plays carelessly, by passing if his first guess is correct, all three players are equally likely to be eliminated. (If Alfred passes after a correct guess, there are two right and one wrong answers left; whoever would select the wrong answer misses out.) If Chips makes the mistake of opting to play, he has a one in four chance of being eliminated. As he dwells on his options, and realizes he is in Alfred's hands, Chips might bewail his poor play in earlier rounds. Had he played aggressively earlier, and got ahead of either opponent for a tie-break, he would have no need to rely on Alfred; he simply plays until the game ends.

This fable illustrates how ignoring the likelihood of early wipeouts may be useful to survive any tie-break. However large a fortune you accumulate, the game's framework almost ensures you will be wiped out several times, no matter how cautious you are. In the final round, you should be aware of the possibility that you can act in the way Alfred was advised to act—he was quite indifferent to whether his answers were right or wrong. It was a vital factor that Betty held zero; if she had a mere £10, Alfred would need to act differently. Near the end of the final game, swift thinking and cold logic are essential.

At an earlier stage, there are so many possible alternative paths to the end of the game that identifying the best decision, play or pass, is far more complex. But this *Wipeout* elimination round is a splendid prototype problem for assessing how to set about making decisions when there are several stages to go through. The method is known as *backwards induction*. To illustrate it, we will use this method in the simpler case of just two players, Angela and Brian, but otherwise using the same *Wipeout* rules. Any money won is irrelevant, the sole object being to eliminate the opponent. We take the final correct object to be worth £100, the previous one £90, and so on in steps of £10.

We want to find Angela's winning chances, and her best tactics, from *any* 'position' in the game, assuming both are guessing at random. Each position can be described by a list of four numbers, (a, b, c, d), where:

- a is the number of correct answers left;
- b is the number of incorrect answers left;
- £c is Angela's current fortune;
- £d is Brian's current fortune.

So $(3, 2, 70, 60)$ means there are five objects left, three correct and two incorrect. The next correct object is worth £80. Angela currently holds £70, Brian has £60. Suppose Angela has the choice to play or to pass. Which should she select, and what is the chance she eliminates Brian?

As soon as either the final correct object, or the final wrong object, has been named, the game ends. For example, at $(2, 0, 120, 0)$ we know Angela has won. At $(0, 1, 90, 100)$ Brian wins, and we shall need to use the tie-break to settle $(2, 0, 0, 0)$. These 'end' positions are those in which the value of either a or b is zero; each end position can be given a label, which is just the *probability* that Angela wins. Where Angela is certain to win, this label is 1, where Brian wins the label is 0, and if the end position needs the tie-break rule, the label is p, the probability Angela wins the tie-break. The three positions we have mentioned will have labels 1, 0, and p. Our aim is to find the labels for any general position, such as $(3, 2, 70, 60)$, and the steps the contestants should take to maximise their own winning chances.

One possible path from $(3, 2, 70, 60)$, with Angela to play, starts with an initial wrong guess. This happens two-fifths of the time, and we then move to $(3, 1, 0, 60)$ with Brian to play. Three times out of four he will guess correctly, and we reach $(2, 1, 0, 140)$, from which he has a choice. If he decides to play again, he guesses wrongly one time in three, leading to the end position (2, 0, 0, 0).

$$ (3, 2, 70, 60) \xrightarrow{\boxed{2/5}} (3, 1, 0, 60) \xrightarrow{\boxed{3/4}} (2, 1, 0, 140) \xrightarrow{\boxed{1/3}} (2, 0, 0, 0) $$

The key to finding the best tactics is the realization that, from any current position, we could, in principle, list *all* possible sequences of future plays. We must soon reach some end position, because after each guess, one fewer object is left; since we know the label for any conceivable end position, *we just work backwards*. The first step is to see what happens when there are just *two* objects left, one being correct and the other incorrect. When Angela (or Brian) plays, s/he will win £100 half the time, and be wiped out half the time, but some end position will certainly be reached. Given their current fortunes, there are just two possible end positions, each equally likely, so Angela's chance of winning is the average of her chances at these two end positions. If either player has the choice of play or pass, s/he should work out Angela's winning chance for each alternative, and make the better choice.

Take the example of the position (1, 1, 0, 90). If Angela plays, we can show the two alternatives, with the labels for the end positions, as

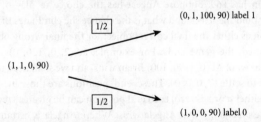

Hence (1, 1, 0, 90) has label '1/2', that being the average of the labels at the two equally likely end positions. If Brian plays, the two possibilities are

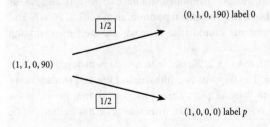

This time, $(1, 1, 0, 90)$ will have label '$p/2$', the average of p and 0. If Brian has a choice to pass or to play, he chooses the alternative whose value is smaller (the label is Angela's winning chance).

In this way, given any position and two objects left, we can label it with Angela's winning probability. Suppose now there are three objects left, one of them correct and two incorrect. After the next play (whoever makes it), one-third of the time we reach an end position with no correct and two incorrect objects left, and two-thirds of the time we have one of each sort left. *In either case, we have already found Angela's winning probability.* Thus we find her overall winning probability by calculating the average of these probabilities, with respective weights 1/3 and 2/3. If either player has the choice to play or to pass, s/he should compute the chances that Angela wins under each alternative, and select the more favourable option. The last step is to label the position with Angela's chance of winning.

Try a problem

Should Angela play or pass from $(1, 2, 170, 0)$?

Suppose first she decides to play. Then we can represent what happens after her guess as

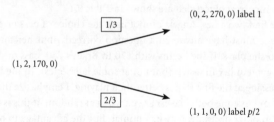

If she plays, her probability of winning is $(1/3) \times (1) + (2/3) \times (p/2) = (1 + p)/3$.

Similarly, suppose she decides to pass, so that Brian will now play.

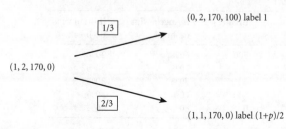

Angela's winning chance is now $(1/3) \times (1) + (2/3) \times (1 + p)/2 = (2 + p)/3$.

Whatever the value of p, this second chance is always larger than the first, so she should pass. The position $(1, 2, 170, 0)$ has label $(2 + p)/3$, if Angela has a choice between play and pass.

Similarly, we make the corresponding calculations for the positions with two correct and one incorrect objects, and give the appropriate label. This would deal with all positions that have at most three objects. Then move on to where there are four left: either three correct and one incorrect, or two correct and two incorrect, or one correct and three incorrect. This leads to Angela's winning chances for every start point with at most four objects left. Our original problem has a start point with five objects (three correct, two incorrect), so we could now deduce Angela's winning chances, since we have found her winning chances for all the positions she could reach in one step. Much attention to detail is required—this is just the sort of problem to give to an adept computer programmer—so I will simply state Angela's winning chances without the intervening working. With optimal play on both parts, the advantage of being able to win via the tie-break rule turns out to be *40 percentage points*, whatever their initial fortunes. Angela's winning chances for various starting fortunes, when there are three correct and two incorrect objects left are shown in Table 9.1.

In the first two cases, Angela cannot have the choice of pass or play (to do so, she must have answered a question correctly, but her fortune is zero); so she plays. If she begins with £70 to Brian's £60, pass or play are equally good at her first step (but not at some later ones). In the last two cases, passing at the first step is better than playing. I emphasize that these calculations assume both players are guessing at random: if this is not the case, then other decisions may be optimal. But the advantage to being in the position to be the winner of any tie-break is considerable. Do not pass

Table 9.1

Angela(£)	Brian(£)	If Angela wins tie-break	If Brian wins tie-break
0	500	60%	20%
0	0	70%	30%
70	60	70%	30%
70	500	70%	30%
130	0	80%	40%

readily in the early stages, seek to build up a healthy balance of correct answers.

I have no expectation that TV *Wipeout* contestants will swot up their backwards induction before their 15 minutes of fame. Phrases about nuts and sledgehammers are apposite. My aim has been to illustrate how that technique could be used in a serious problem where a sequence of decisions is needed, and chance plays a role. Perhaps you are ten years from retirement, but must choose how to invest a lump sum each year to provide for your eventual pension. For all possible values of your accumulated fund at the end of year 9, you would know whether to opt for the risky or the safe path in year 10. Your decision at the end of year 8 will depend on which of the year 9 possibilities seem most attractive and achievable, and so on back to your initial commitment now.

It is rather like deciding how to cross a river that has many stepping stones at various distances from the far bank, up and down the stream. For any stone close to the far bank, you can assess your chance of a successful final step. Some of these stones are more attractive than others. Two steps away, you would like to be on a stone from which you can move to a good stone for the final step. Work backwards to plan how to begin.

The Showcase showdown

At the climax of the game show *The Price is Right*, three contestants, Annie, Boris, and Candy, stand in front of the host. One of them will win the right to go on to try to win the Showcase, a cavalcade of prizes. The decision is made using a giant wheel, divided into segments, on which the numbers 5, 10, 15, 20, … , 100 can be seen. They all appear once, in some random order. Annie goes first, spins the wheel, and sees her score. If she is happy with that score, she stops, and the spotlight moves to Boris. However, Annie may decide to make a second spin. If she does so, her score is the total of the two spins, and that ends her turn. Boris and Candy have the same choices, and all three of them watch the scores unfold. The object is to make a score as close as possible to 100, *but without going over 100*. Anyone whose score exceeds 100 is automatically eliminated. If Candy is lucky enough to see both Annie and Boris make this error, she wins whatever the outcome of her first spin. If there is a tie, those involved have a spin-off to see who scores most on one spin.

Should you stick with your score after one spin, or should you take a second one? When should you press on, if your score ties that of a previous player? How advantageous is it to be in Candy's place, to be able to see the two previous scores? Who will win this showdown?

These questions are similar to the choices faced by contestants in *Wipeout*, but the calculation of the best policy is easier. The first step is to consider the case when there are just two players, Annie and Boris. To work out what Annie should do, we first have to see how Boris would respond to the various scores she could make.

All is clear if Annie's score exceeds 100: she is eliminated, and Boris wins. Otherwise, her score is in the range from 5 to 100, and Boris makes his first spin. Most of the time, his action is obvious: if his first spin is less than her score, he must spin again, and if his score exceeds hers, he stops. But suppose his score is equal to her score. His mind should rehearse the following argument: 'If I stop, we move to a spin-off, and then each of us is equally likely to win. So I should spin again if that makes my chance exceed 50%. There are 20 possible scores. If our common total is 45 or less there are more scores that favour me than favour her, so I should spin; with a score of 55 or more, I am more likely than not to bust. With a score of 50, I am equally likely to win or bust, so it makes no difference. I'll spin on 50 or less, and go to the tie-break with 55 or more.'

Annie assumes Boris would use his best strategy. She considers her two alternatives: to stop or to spin again. She will choose whichever gives the higher chance of winning. As her initial score increases through 5, 10, 15, . . . , 100, so her chance of winning (if she stops) steadily increases. But if she decides to spin again, she might overshoot and lose straight away. This factor leads to the interesting and helpful result that as her initial score increases, so her chance of winning *if she spins again* steadily decreases. There will be some point where these two chances cross over. After several lines of arithmetic, that point can be calculated to be at a score of 50. So long as her initial score is 50 or less, she should spin again, but with a score of 55 or more, she should stop. Her chance of winning if she stops with 55 is only 29%, but if she spins, that would slip to 28%. Overall, if she uses this strategy, she wins about 46% of two-person contests, and loses 54%.

To see what happens with three players, we again begin with the last player, Candy. She is competing against the better of Annie and Boris. The only time she needs any advice is when her first spin puts her in a tie with the current leader. Suppose she reaches this point, and ties with just one other person: then she is in exactly the same position as Boris when he reached a tie with Annie. So Candy should go to a spin-off if the tie score is

55 or more, but spin again at 50 or less. But suppose both Annie and Boris have this same score, so that a spin-off would be a three-way contest. She argues: 'If there is a spin-off, all of us are equal, so my chance is one in three. So I should spin again whenever my chance of victory exceeds that. If the common total is 65, there are seven spins (those from 5 to 35) out of the 20 possible where I win. But if the common score is 70, there are only six spins where I win. Since seven out of 20 is better than one in three, whereas six out of 20 is worse than one in three, I should spin again at a total of 65 or less, and trust in the tie-break with 70 or more.'

The focus moves back to Boris. He has gone through the previous work, and knows how Candy will play. If his first spin is behind Annie, he obviously spins again. If his first spin takes him ahead of Annie, he is now in a two-player game with Candy, which has the same format as the two-player game we have already dealt with. What should he do if his first spin takes him equal to Annie?

The larger this score, the more likely he does better to stop. As before, if he spins again, he might overshoot, but this time he has to think about the possibility of a three-way spin-off. When the dust has settled, the magic number is 70: whenever he and Annie have a score of 70 or more he should stop, if their common total is 65 or less he should spin again.

Finally, consider Annie. She has to work out the whole of the above to be sure of acting optimally, so she had better have done it earlier or the TV audience will get impatient. Her arithmetic is indeed very detailed, but leads to a simple criterion: if her first spin is 65 or less, she should spin again, but if it is 70 or more she does best to stop. Stopping with a score of 70, her chance is about 21.5%, but that would drop to 17% if she were to spin again.

We now know what all of them should do. It is an advantage to be in Candy's position, but not so great as you might think. If all three accept the advice given, Annie's chance is just under 31%, Boris wins nearly 33% of games, and Candy goes for the Showcase over 36% of the time.

I have suppressed nearly all the details, and for a very good reason: they would occupy several pages. If you want to be convinced by a full printed argument, read the article by P. R. Coe and W. Butterworth that is listed in the references. The gory detail is in working out the probabilities for all the possible positions in the game. Recall that one of the assumptions in this game is that all the scores on the wheel have the same chance of occurring. If some numbers came up more often than others, the arithmetic—already forbidding—would be ten times worse.

Blockbusters

This is a quiz game, usually played by teams and individuals aged 16–18. Figure 9.1 shows a *Blockbusters* board, with $m = 4$ rows and $n = 5$ columns; this is the format used in the TV game, but any number of rows and columns could be used. Each hexagon is labelled by a different letter, which helps us to identify the different positions, but in the quiz game it is also the first letter of the correct answer. For different games, the positions of the letters (and the letters themselves) will change.

A snatch in the middle of the game might run:

Host: 'Well done, Richard, that's correct, so it's now your turn to choose a letter.'
Richard: 'I'll go for a P please Bob.' {Traditional titter from audience.}
Host: 'What P is the element whose chemical symbol is K?'
(Sound of buzzers.) Host: 'I think Viresh got in first; yes Viresh?'
Viresh: 'Potassium.'
Host: 'Correct; a White gets filled in this time. Your choice, Viresh,' etc.

To begin with, the board is blank, apart from the labelling letters. Whenever a question is correctly answered, the corresponding hexagon is coloured Blue or White, as appropriate. The Blue team has two players, and is seeking to answer questions that will give a continuous linked path across the board, from left to right. The White team has just one player, who is similarly seeking a continuous linked path from top to bottom. The order in which the questions are answered does not matter, and the game ends as soon as either team has achieved its objective. The first hexagon to be contested is chosen at random, and after this, the team that wins any hexagon has the right to select the next one to be fought over. If neither

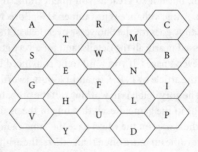

Fig. 9.1 A 4×5 *Blockbusters* board.

team is successful with a particular question, another question for the same hexagon follows.

There is a prize of £5 for each question correctly answered, but the main prize comes from winning a contest. Each match is played as a 'best-of-three' series between the single player and his or her two opponents, and the winning team gets to play another contest (but up to some maximum). The host insists on a rule to prevent a game being dragged out to an unnecessary and perhaps humiliating length: if there is a potential winning hexagon available to the team making the choice, it must select such a hexagon.

Every contest must end with a winner, there cannot be a draw. To see this, imagine each of the 20 hexagons coloured in as either White or as Blue. If there is a linked Blue path from left to right, no matter how twisted, this will prevent any linked White path from top to bottom—such a White path is blocked by the Blue one. And if there is no Blue path from left to right, there must be a White one from top to bottom: if there were no such White path, then Blue would have no obstacle to its left–right journey. (Think of the analogy of a Blue river seeking to flow from left to right, and an attempt to build a White dam from top to bottom; either the dam stops the river, or the river gets through—and not both.) This argument applies to a *Blockbusters* board of any size, not just with four rows and five columns.

Is the game fair to both teams? One team has two players, the other team has only one player, so one team has twice the chance of the other of buzzing in first with the right answer. In compensation, the two-person team has to travel across the board, a minimum distance of five steps, and maybe six or seven if White gets some good blocks in. The one-person team could have a winning path with only four correct answers. So suppose each of the three contestants is equally skilful, with each player having the same chance of winning any hexagon: then for any hexagon, the Blue team's winning chance is 2/3, and the White player has a chance of 1/3. How does this translate into winning a board, and then into a best-of-three contest?

Although the actual TV game ends as soon as either a Blue path or a White path has been created, the best way to find the teams' winning chances is to pretend the game continues until all 20 hexagons have been filled. This cannot change the identity of the winners, since the original winning path remains a winning path no matter how the rest of the hexagons are coloured. The advantage of this extension is that now we can suppose that all games are contested over the same number of questions.

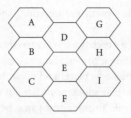

Fig. 9.2 A 3 × 3 *Blockbusters* board.

This step is similar to one made in the last chapter, and for the same reason.

A 4 × 5 board enables paths to twist around so that sometimes it is not immediately obvious whether the game has been won by Blue or White. Before we find the chance for each team on this board, look at the simpler case when there are three rows and three columns, as in Figure 9.2.

For this possible game, a clear advantage to Blue can be identified. Suppose there are just three Blue hexagons, the minimum number that can give a left–right path. From the figure, there are plainly nine left–right paths of length three, namely:

ADG ADH BDG BDH BEH BEI CEH CEI CFI.

However, there are only seven top–bottom paths of length three, i.e.

ABC DBC DEC DEF DEI DHI GHI.

So if Blue gets exactly three correct answers out of nine, Blue's chance of having a winning path is better than White's when White holds three correct answers! The same holds with four correct answers: there are 46 winning Blue paths, but only 39 winning White paths. With five Blues, there are 85 colourings in which Blue wins; with five Whites, there are only 80 colourings where White wins. With six correct, Blue wins 77 times, while White wins just 75 times. At every turn, the game's layout favours Blue! The same phenomenon occurs in the 4 × 5 board used in the TV game.

Suppose a game between two equally skilled players were played on the 3 × 3 board of Figure 9.2. How does this asymmetry in the layout translate into Blue's chance of victory? The assumption that the two players are equally skilled means that each hexagon is equally likely to be coloured Blue or White. There are two ways of colouring each of nine hexagons, so the total number of ways of colouring the whole board is 2 × 2 × 2 × ...

$\times$ 2 (nine times) = 512, all equally likely. Of these colourings 247 have a White path from top to bottom, 265 have a Blue path from left to right, so Blue's winning chance is 265/512 = 52%. This shows that even with two players of equal skill, Blue wins 52% of the time, and White just 48%! Should you ever have the opportunity to play in such a contest, wheedle your way into having the left–right path. In a 5 $\times$ 5 board, Blue's corresponding winning chance increases to 53.5%.

Return to the TV game, where we want to find White's winning chance when White wins any hexagon one time in three. Sometimes, for example, exactly nine of the 20 hexagons will be coloured White, with the other 11 coloured Blue. When this occurs, some of these colourings will have a Blue path, the rest will have a White one. The key step is to *count* how many colourings fall into the two categories. This is not easy, because of the myriad ways a path can sneak around, but it is possible to write a computer programme that will identify whether any given colouring has a White path. Then we just generate each colouring that has nine White and 11 Blue in turn, and use this programme to see whether or not there is a White path. To give some idea of the scale of the problem, I simply report that of the 167 960 possible colourings of the standard *Blockbusters* diagram that have nine White and 11 Blue hexagons, 70 964 of them have a White path, and the other 96 996 have a Blue path. But since White wins any hexagon with probability 1/3, while Blue has probability 2/3, the probability of any one of these colourings is

$$(1/3)^9 \times (2/3)^{11}.$$

That means that the total probability of having a game with nine White and 11 Blue hexagons, that ends in a win for White, is 70 964 $\times$ $(1/3)^9$ $\times$ $(2/3)^{11}$ = 0.0417.

We now do the same sort of calculation for colourings other than with nine White. Plainly, if there are fewer than four Whites, a White path is impossible, and if there are 16 or more Whites, a White path is certain. The details are given in my paper, listed in the references. Adding up all the corresponding probabilities, the total chance for White is a tiny 0.1827! And since to win the contest, White must win two out of three games, his overall chance is even less than this. You might well have guessed that White had some disadvantage by having to compete against two opponents, despite the shorter White paths, but to find the disadvantage is as large as this is alarming.

The analysis of Figure 9.2 showed that going across the board is advantageous. Perhaps the two-player v. one-player format has some particular appeal, but the disadvantage to White is accentuated by the orientation chosen: instead of a 4 × 5 board, suppose it were changed to be five rows and four columns, with White to take the shorter left–right path. White is still at a large disadvantage, because with three equally skilled players, the single White player wins only 1/3 of the points. However, this change would increase his winning chance modestly, to 0.2245. So long as TV screens are wider than they are high, we are unlikely to see this change. A simpler proposal: make the game 3 × 5—White's chance of winning a board is increased to 0.3222, which seems reasonably fair with three players.

In the last chapter we found a formula to convert White's chance of winning one game into the chance of winning a best of three. When the chance for a single game is around 20%, his chance of victory over the series is about 10%. So salute any White individual who wins several victories. That person's speed of thought and reaction, and volume of trivia knowledge, are demonstrably well in excess of that shown by his or her opponents.

Some hexagons are more important than others, in the sense that if you win them, this opens up more winning paths. In Figure 9.1, hexagon F is in the most useful position, with Y and D being least helpful. This observation is mainly of academic interest, as a team's chance of winning any particular hexagon is unaffected by any importance it has, and it is never a disadvantage to hold a given hexagon. But remember that you can select whichever hexagon you wish, so long as you are not on the verge of completing a path. It makes sense to try to pick up as many hexagons as you can (each one is worth money), as the order in which the hexagons are contested makes no difference at all to either team's winning chances. So the recommended strategy is for both teams to dodge around the board when it is their turn to select a hexagon, as this simply increases the total prize money either team wins. Do not attempt to construct a short path, but be careful not to choose a hexagon that would allow the other team to win, or to get on the verge of victory. Hexagons in the positions labelled Y, D, R, A, and C will often fill this bill.

Postscript

A *Blockbusters* board is similar to a board in the game of Hex, invented independently by Piet Hein in 1942 and John Nash in 1948. Figure 9.3

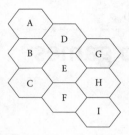

Fig. 9.3 A 3 × 3 Hex board.

shows a 3 × 3 Hex board, which can be compared to the *Blockbusters* format of Figure 9.2.

In Hex (normally played on a *much* larger board than that of Figure 9.3), White and Blue play alternately. Each chooses one hexagon to colour; Blue seeks a linked left–right path of Blues, White looks for a White path from top to bottom. Unlike *Blockbusters*, there is no natural advantage either to left–right or to top–bottom. For example, there are just eight Blue paths of length three, namely

ADG   BDG   BEG   BEH   CEG   CEH   CFH   CFI,

and also just eight White paths of length three,

ABC   DBC   DEC   DEF   GEC   GEF   GHF   GHI.

The same is true when either Blue or White holds any other number of hexagons, on any Hex arrangement with the same number of rows and columns. Just as in *Blockbusters*, every game ends in a victory for one side or the other, and for the same reason. In the game as specified by Nash or Hein, it can be proved that there is a winning strategy for whoever has first move, provided the numbers of rows and columns are the same. But *finding* a winning strategy, for even a moderately large board, appears quite difficult.

It is very easy to construct a fair game, along *Blockbuster* quiz lines, but using a Hex board. All you need is the same number of rows and columns, and teams of equal size. But until TV screens are made in the shape of a rhombus with one angle 45%, rather than a rectangle, such a game would be for home amusement only.

CHAPTER 10

Casino games

The main purpose of the Betting and Gaming Act, 1960, was to provide a legal framework for off-course betting on horse-racing, but it also led to the opening of some UK casinos. The 1968 Gaming Act brought stricter controls over casino operation, including rules on where they may be sited. A Gaming Board determines what games are permitted, advises the local magistrates who actually grant the licences, and carries out checks on the character of employees. The annual 'drop'—the amount of money exchanged for gambling chips—is some £2.6 billion, of which £470 million stays with the casinos. London, with more than 20 casinos, accounts for 80% of the UK turnover.

Legislators have had an ambivalent attitude towards commercial gambling: it happens, but should not be seen to happen. The 1960 Act recognized the inevitability of gambling, and sought to regulate it, tax it, and hide it. Advertising by casinos and betting shops was not permitted, and demand for gambling was not to be stimulated. The casino operators have used the advent of the National Lottery to argue successfully for a relaxation of these restrictions, but ambivalence remains. National advertising is to be allowed, but not in publications that generally promote casinos. The number of gaming tables can be mentioned, but not the size of prizes. Soon the Yellow Pages will have an entry between cash registers and casks, and tourists will use post or fax to apply for casino membership. A casino may serve meals, but may not offer separate entertainment: no pop concerts, no poetry readings. Contrast Las Vegas, where to hear Tom Jones you must make your way to the furthest hall in the house, past fruit machines and gaming tables on both entry and exit.

Casinos make their profits mainly through the built-in house advantages in the various games offered. To many who visit the casino, the size of the house edge is irrelevant, so long as it is not exorbitant. Gambling losses are a fair exchange for the social occasion, the chance to meet friends, the regular night out. This chapter is also for the punter interested in these odds, who seeks to discover which games have a smaller house edge, whether there are ways to exercise skill, and how best either to spin out your funds, or to maximize your chance of reaching a desired goal. The games offered, and their odds, are those that apply to casinos on the UK mainland; the arm of the Gaming Board does not extend out to sea. Roulette wheels on ships may be identical to those on land, but the odds offered are usually worse.

Most of this chapter studies roulette and casino stud poker. I hope not only to be informative about various approaches you can take in casinos, and your chances in these games, but also to illustrate ways of working in probability. One point that arises in the description of the poker game is the interaction between your hand and the dealer's possible holdings. If the two hands were independent, so that your hand had no effect on what the dealer could hold, finding the answers would be much easier. But if you hold three Kings, the dealer cannot also do so. There are nearly 2.6 million possible hands from a full deck of cards, but when your five cards are removed, that leaves only 1.5 million possible dealer hands. Your own hand prevents the dealer having over 40% of theoretically possible hands.

Despite this impact your hand has, it transpires that the decisions you make, your chance of winning, and the house edge are very close to what they would be if the two hands were dealt independently from different packs. *You cannot know this for sure before you do the sums.* I regret that the calculations are complex and unattractive: you are free to skip those few details I have not suppressed. But we do not live on plovers' eggs and truffles with vintage champagne all the time. Some days, only plain greens and tap water are available.

Roulette

This is by far the most popular casino pastime, taking about 60% of total stakes. The standard UK roulette wheel has 37 numbers, 0 to 36. The number zero is coloured green, the other numbers are equally divided between red and black, as shown in Figure 10.1. The gambler's cash buys an equivalent number of coloured chips. There are no disputes about whose bets have won, as everyone playing at the same table has different colours. Within the same casino, some tables may offer very cheap bets, while other tables play at higher stakes. There is a maximum stake, typically about 100 times as big as the minimum, for any bet permitted.

Fig. 10.1 A standard roulette wheel.

Figure 10.2 shows the range of possible bets, and the associated odds paid out to winners. We shall assume that all 37 outcomes are equally likely: casinos have a powerful incentive to meet this ideal, as their advantage is so small that any detectable bias might tip the odds in favour of alert gamblers. To find the size of the house edge, we discover what will happen, on average.

For any bet other than those that pay out even money, the average return is 36 units for every 37 staked. However punters combine bets, miss turns, change the size of their stake, or vary their play in any way at all, this average remains constant. One unit is lost for every 37 staked, a house advantage of 2.7%.

But for the bets on Red, Black, etc. that pay out even money, the house advantage is half as much. This is because of the rule for these bets that when zero is spun, the house takes half the stake, and the other half is returned to the punter. On these bets, the house edge is reduced to 1.35%. For simplicity, we will use the phrase 'a bet on Red' to denote any one of the six bets that pay out even money.

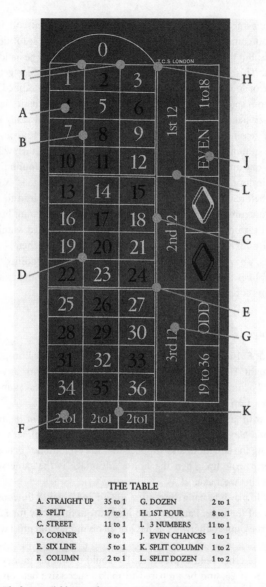

THE TABLE

A. STRAIGHT UP	35 to 1	G. DOZEN	2 to 1
B. SPLIT	17 to 1	H. 1ST FOUR	8 to 1
C. STREET	11 to 1	I. 3 NUMBERS	11 to 1
D. CORNER	8 to 1	J. EVEN CHANCES	1 to 1
E. SIX LINE	5 to 1	K. SPLIT COLUMN	1 to 2
F. COLUMN	2 to 1	L. SPLIT DOZEN	1 to 2

Fig. 10.2 The bets available at roulette.

This rule means that when a gambler stakes two units on Red, the normal outcome is that she will either lose her stake, or see it double to four units. But the one time in 37 that zero is spun, she will be handed back one unit. This gives three possible outcomes—the gambler gets either zero, one or four units. But suppose the zero rule were modified, so that when zero is spun, anyone who bets on Red is given one chance in four to double her stake, otherwise she loses all of it. This could easily be done either by tossing two coins to see whether both were Heads, or by an auxiliary spin of the roulette wheel. Plainly, her *average* return when zero is spun is one unit (she has one chance in four of receiving four units, three chances in four of receiving nothing), so the casino's edge is exactly the same as before. But now every stake of two units would lead to just *two* possible outcomes, four units or nothing. Taking into account how often zero will be spun, this change is equivalent to the bets on Red winning with probability 73/148, and losing with probability 75/148. When we discuss these bets, we will act as though casinos behave in this modified fashion. This simplifies the analysis considerably, without making material difference to any conclusion worth drawing.

Reaching a goal

Paul needs £216 for an air ticket to see his team play in the European Cup Final tonight. He has half that amount, £108, but might just as well have nothing. To him, £216 is Nirvana, but even £215 is as useless as an expired credit card. Can his local casino help?

Faced with any similar problem, there is one maxim: *in unfavourable games, bold play is best, timid play is worst.* To advise Paul, let's begin by simplifying the situation, and suppose that the even money bets also lose when zero turns up. Then the house advantage is the same whatever individual bet we make.

One bold play is to bang the entire £108 on Red.Then 18 times out of 37, or 48.6% of the time, Paul can cash his chips immediately for £216, and travel to the game. Otherwise, he has lost all his money and must watch the match on TV.

There are other bold approaches. He could divide his money into 18 equal piles of £6, and be prepared to makes successive bets on a single number until he has either run out of cash, or one bet has succeeded. Single numbers pay at 35:1, so one win is enough. To calculate his chance of success, we first find the chance that all his bets lose. Any bet loses 36 times in 37, so the chance they all lose is $(36/37)^{18} = 0.61$. For this strategy, the chance that at least one bet wins is 39%, rather worse than the single bet on Red.

Paul has an alternative way to use the 35:1 payout on single number bets. He already has £108, so if he uses just £4 on the first bet (£3 is not quite enough), a win would take him to his target. Indeed, so long as he has at least £76, a winning bet using £4 gains enough; with less than this, he should stake £5, and when he has less than £41 he must increase each bet to £6. In this way, he can plan to make a total of 22 bets if necessary (nine of £4, seven of £5, six of £6); if any of them win, he has reached his target, otherwise he is reduced to £1. The chance they all lose is $(36/37)^{22}$, about 54.7%, so he succeeds only 45.3% of the time. His last-ditch hope is to win a 5:1 bet with the final £1, and then a 35:1 bet using £6: this takes his chance up to 45.7% only. Overall, this is better than trying the 18 bets of £6, but not so good as one big bet on Red.

One final try: place his whole fortune on the odds-on bet of a split column. If that wins, he has £162, so he can now place £54 on Red; if that wins, victory; if it loses, he is back to £108 and can begin to repeat this sequence. This strategy wins only 47.3% of the time. All of these alternative attempts are inferior to the original suggestion of a *single bet* on one of the even money chances. And in reality, that bet is even more favourable than we have supposed, because of the zero rule on the bets on Red. Taking that into account, Paul has a 49.3% chance of getting to see the match. His stay in the casino will be short, but he has no interest in the extended pleasure of watching his fortune fluctuate; it is Gatwick, or bust.

The numbers used were chosen to make the arithmetic easy, but bold play means *making as few bets as possible*. Take the case when Paul has £24, instead of £108, but still yearns for £216. His expectations are low, but he still seeks his best chance. One bold play is to place all his money on an 8:1 corner bet, which succeeds if any one of four numbers is spun. His chance is 4/37, nearly 11%. An alternative is to split his money into two equal amounts, planning to make two split bets at 17:1 on successive spins. Here he succeeds only 10.5% of the time. No other bet, that does not use the bets on Red, can do better than the single corner bet.

Starting with £24, Paul's bold way to use the Red bets is to risk his entire stake when he has up to £108, hoping to double up, and to use just enough to reach his target when he has more than £108. Because of the smaller house edge on the Red bets, the comparison with the corner bet is not clear cut. It turns out that with £24 he has to bet so often that this advantage is eroded, and the corner bet is still superior. If he had £54, he starts nearer his target, and there is no real difference between intending to make two Red bets on successive spins, and covering nine single numbers with £6 each on a single spin. In either case, his chance of quadrupling his money is just over 24%.

Will you be ahead when you quit?

Paul's brother Michael also has £108, but he will certainly watch the match on television. Providing his money lasts out, he will leave the casino at six o'clock, his visit being merely for recreation. What is the probability he is winning when he calls it a day?

The answer will depend on how many bets he makes, and what they are. Suppose he confines himself to one type of bet, and stakes £1 every time. For ease of comparison, he will not use the bets on Red which are marginally more favourable, but will bet on some block of m numbers, where m is either 1, 2, 3, 4, 6, 12, or 24, as shown in Figure 10.2. Whichever choice he makes, the average result is a loss of £(1/37) each spin, so if he had time for 370 bets, his average loss would be £10.

If he bets only on the single numbers, the variability about this average will be more than if he uses one of the other bets. The variability will be least on the split column bet of 24 numbers. The greater the variability of the outcomes, the greater the chance that the actual result will be well away from the average, either bigger or smaller. Since, on average, he makes a loss, his best chance of being ahead when he stops is when the variability is large. He should go for the 35:1 payout on single numbers. This choice also increases the chance that his loss will be much worse than average, but this is not the issue. Michael's sole aim is to maximize the chance that he is ahead when the clock strikes six.

Join Michael in the gaming room as he makes his bets, and follow his chances of being ahead as the session continues. It is very likely that he will begin with a series of losses, and so fall behind, but when he wins, he increases his fortune by £35. This means that he will suddenly move ahead if he is lucky on *any* of the first 35 spins; only when all of them fail will he then be losing. From an initial chance of 1/37 after the first spin, his chance of being ahead steadily increases up to and including spin number 35. The chance that all the first 35 spins lose is $(36/37)^{35} \approx 0.3833$, which gives Michael a 62% chance of being ahead after the 35th spin.

But to be ahead after 37 spins, he would need to have struck lucky at least *twice* in that period. In 37 spins, the chance of no wins at all is $(36/37)^{37}$, and the chance of exactly one win is $(36/37)^{36}$. That means that his chance of two or more wins by after spin 37 is $1 - (36/37)^{37} - (36/37)^{36}$, a dramatic drop to under 27%. But from this time onwards, up to and including 71 spins, he will be ahead if he has two or more wins in total, so his chances increase over this time. After exactly 71 spins, his chance has moved up to nearly 58%.

In the same fashion, of course, he needs to have three or more wins to be ahead after 73 spins, so his chance of being ahead at that instant plummets again, this time to below 32%. From here up to spin 107, a total of three wins is enough, so his chances increase, but plunge again at spin 109. (I have carefully skipped over spins 36, 72, 108, etc., because at these times it is possible that he is exactly back where he started; it then becomes an exercise in semantics to say whether being ahead does or does not include being level. To avoid this diversion, stick to numbers of spins that are not multiples of 36.)

The picture is clear. Michael's chances of being ahead at various times in the session follow a saw-tooth pattern. There is an initial steady increase followed by a dramatic drop either side of 36 spins. It increases again, but to not quite the previous level, with another drop around 72 spins, followed by a similar rise for 35 spins, and a drop either side of spin 108, and so on. A real saw manufactured to this exact pattern would be rather inefficient, as each successive maximum, just before the drop, is a little lower than before. When six o'clock arrives, the chance that Michael is ahead depends heavily on *where* in these cycles of 36 spins fate has placed him. If the total number of spins is just under a multiple of 36, his chance of being ahead when he quits may be quite high, but if this number is just above a multiple of 36, it will be rather lower. At a rate of 90 spins per hour, Michael would make about 180 stakes in a two-hour session. To be in the lead after 179 spins, five wins would be enough, and the chance of this or better is over 53%. But to be ahead after 181 spins, he would need at least six wins, and the chance of this is only 37%.

If his sole object were to give himself the largest possible chance of leaving the casino ahead, after a series of bets on single numbers, he should ignore the clock, and plan to make no more than 35 spins—one win then suffices, a 62% chance. More plausibly, he plans to stay rather longer, perhaps for a time that would give about 180 spins. In that case, to maximize his chance of being ahead at the end, he should plan to make 179 bets.

Actually, some of these calculations have a small flaw, since we took Michael's initial capital to be £108. We assumed he would be capable of making about 180 plays, but he might have so many initial losses that his capital is wiped out before his planned exit. His limited capital reduces the probabilities mentioned by about 0.5 percentage points. But the comparison between quitting after 179, or 181 plays, remains unaltered, and if he has at least £181 capital, the figures stand. On the other hand, if Michael had only £54 to start with, there is a chance of nearly one in four that he loses the whole lot before any win at all—and casinos do not allow credit (yet).

Instead of betting on single numbers, Michael might try his luck with blocks of 2, 3, 4, 6, or 12 numbers. The overall saw-tooth pattern of his chance of being ahead occurs again, but this time the sudden drops are around multiples of 18, 12, 9, 6, or 3 spins, instead of multiples of 36 spins. The outcomes are less extreme, the sudden drops less dramatic. Suppose he concentrated on the 11:1 payout for betting on a street, a row of three numbers. After 179 or 181 spins, his chances of being ahead are 49% and 40% respectively. If he used the 2:1 payout on a column of 12 numbers, the comparative figures are 40.5% and 38.5%.

Could Michael use the bets on Red, with their smaller house edge, to increase the chance that he walks away richer? Not unless he plans to make thousands of spins. Over a single session, where he makes at most several hundred bets, he is more likely to be ahead at the end if he sticks to the 35:1 payouts. After 179 bets there, he is ahead 53% of the time; with any number of bets on Red, his chance of being ahead can never exceed 50%. Sticking to Red, his chance after 35 bets is 47%, after 179 bets it is 44%. On average, Michael does better with bets on Red, but the outcomes are not sufficiently variable to give him as good a prospect of walking out richer at the end.

A flight of fancy

If casinos allowed unlimited stakes and betting on credit, there *would* be a way to be absolutely certain of a profit. All you need do is to bet repeatedly on Red, doubling the stake after each loss. When Red comes up and you win, start the system again. Try this with a few sequences. No matter how many losing bets you begin with, the amount you receive the first time you win is always enough to leave you exactly one unit in profit.

But two fundamental flaws prevent this 'system' being used:

- Casinos do not allow credit, so if you embark on this line, you must have the capacity to stake whatever it takes to continue.

- Even an immensely rich gambler may reach the house limit before his ship comes home.

Suppose that limit is 100 units, and George embarks on the 'system'. His aim is to win just one unit and then walk away. He has a splendid chance of success; he wins his one unit whenever any of the first seven spins are Red. Only when none of them show Red does the house limit interfere, and the chance of seven non-Reds is $(19/37)^7$, which is less than one in 100. Since zero might have been one of these losing bets, he is even better off, so his chance of winning certainly exceeds 99%.

But to dampen your enthusiasm for a strategy that is so overwhelmingly likely to make a profit, contrast your profit of one unit when you win with your loss of 127 units from stakes of 1, 2, 4, ..., 64 when all bets lose. On average, you lose. The combination of a house edge on any bet, and a house limit, is enough to ensure that *every* collection of bets, whether on one spin or on many, is at a disadvantage to the punter. There can be no 'system' where the punter is at an advantage, unless the wheel is flawed. Sorry, folks. If you read Graham Greene's *Loser Takes All*, you might think that a sufficiently expert mathematician will be able to construct a winning system, but you would be wrong. Mathematics shows the exact opposite: *no such system can exist*. (Actually, Greene's 'mathematician' was merely an assistant accountant who styled himself as a mathematician, but even accountants generally know better.)

How to lose

Most casino players manage to lose easily enough without my advice. Tips on losing are seldom sought. But it is not unsound to enter a casino with the intention of losing a definite amount. Anyone misguided enough to start the gambling session intending to quit when they are ahead, or have reached some positive goal, will face the dilemma of the last section. Look ahead to the account of Dostoyevsky's experiences for a warning. To ensure success might require more capital than you have, or a larger stake than the house will permit. Be a pessimist: decide at the start that you will quit when you have managed to lose some definite sum. Every so often, you will massively under-achieve on your ambition. There is then no law that stops you changing your mind, should you find losing too difficult! The advantage of playing until you lose a definite amount is that you have a floor under your losses, and so are in no danger of penury.

One convenient way to work this strategy is through what is known as a (reverse) Labouchere system. To lose (say) 45 units, write down the numbers 1 to 9 in a column (they add up to 45). Then add together the top and bottom numbers, 1 and 9, to form your first stake; place your ten units on Red. If the bet loses, delete the two numbers you used, and move on to the new top and bottom numbers, 2 and 8. But if your first bet wins, make no deletions and write your winnings, ten, at the bottom. The new top and bottom numbers are 1 and 10, so your next bet is 11 units on Red (or Black, of course).

Continue in this fashion, crossing out numbers when you lose, and writing a number in when you win. Stake the sum of the two numbers currently at the top and bottom. If you are ever left with just one number, that is your stake. When you have crossed out all the numbers, both the originals and the write-ins, congratulations. You have achieved your ambition, and lost your 45 units.

But in your efforts to rid yourself of your capital, maybe, just maybe, you find you are not successful, as you are winning too often. Perhaps your system requires a stake greater than the house permits! The only way that can happen is if you are way ahead. You did not sign a binding covenant to play for ever, you are entitled to quit at any time. This system guarantees a loss, eventually, but of not more than 45; keep a watch on your pile of chips, and be willing to change your mind, and quit when you are winning, or it is time for dinner. Your failure to lose could pay for the meal. Of course, the numbers 1 to 9, and their total of 45, can be replaced by any other (positive) amounts to suit your pocket.

It would be foolhardy to play this system the other way round, intending to quit when you have won your target, by interchanging the rules for writing in and crossing out. Lucky old you if you should win, unmitigated disaster otherwise.

The drunkard's walk, or spinning out your fortune

Keeping track of a gambler's fortunes in a casino has strong parallels with watching a drunkard confined to a narrow street. He is so inebriated that his past steps have no influence on which way he goes next, but he may have a small bias one way or the other. His walk will end either with the calamity of a stagnant canal at one end of the street, or the safety of home at the other. Whether he finds disaster or security depends on random chance.

CANAL   PUB   HOME

Take the case where he is *equally likely* to go left or right at any step, and the pub is 50 paces from the canal, and 150 from home. The boxed argument (next page) leads to the following conclusions:

- His respective chances of hitting the canal first, or home first, are 3/4 and 1/4. Since home is three times as far, its chance is one-third as much.

- On average, he takes $50 \times 150 = 7500$ steps before his walk ends. For this type of calculation, just multiply the respective distances to the endpoints.

If there were no house edge in roulette, then betting unit amount each time on the even chances would follow this analysis exactly. The position of the pub is your initial capital, the canal is bankruptcy, and home is when you would quit, if ever you increased your capital to that amount. With no edge, you win half your bets and lose half. The average length of the walk is the average number of bets until your fate is decided.

The more ambitious your target, the less chance you will hit it before you become bankrupt. Not only does your chance decrease, it reduces all the way down to zero. So if you do not set any upper limit at which you would quit, it is certain that you will eventually be bankrupt, *even when there is no edge for the house.*

What happens if you double the amount of your standard bet? In this new currency, your initial capital is just half as big, but then so is the quitting target, and their ratio remains the same. This means that doubling your stake makes no difference at all to your prospect of reaching your goal. The game will tend to be much shorter, of course. In the new currency, both your capital and the amount you want to win are half as much, so on average the game lasts only a *quarter* as long.

Casinos do not offer fair bets, but you and your friend could conduct a fair game of tossing coins. Perhaps you have £1, she has £1000, each toss is played for £1, and the game ends when either of you is bankrupt. This corresponds to you seeking to increase your fortune from £1 to £1001, and so your chance of doing so is a rather depressing 1/1001. The average length of the game is found by multiplying your initial stakes. That means the game lasts on average $1 \times 1000 = 1000$ tosses—perhaps an astonishingly high number. Since half of all games end after the first toss, this

The drunkard's walk

Each step is taken to be the same length, and equally likely to be left or right. The whole street is L steps long, with the canal at position 0, and home at position L. He begins at the pub, n steps from the canal and so $L - n$ steps from home.

When he is k steps from the canal, write $p(k)$ as the chance he gets home before he hits the canal. If he is at the canal, where $k = 0$, it is impossible that he gets home first, so $p(0) = 0$. If he is at home, where $k = L$, he is certain to be home first, so $p(L) = 1$. Look at an intermediate point, k, and see what happens after one step; half the time he moves to $k + 1$, half the time to $k - 1$. So

$$p(k) = \frac{1}{2}\, p(k + 1) + \frac{1}{2}\, p(k-1). \qquad (*)$$

This is true at all intermediate values of k, and we have just the right number of equations for the number of unknowns. The format of $(*)$ is termed a *difference equation*, for which there are standard methods of solution. Omitting the details, the answer is $p(k) = k/L$ (which is very easy to check). So we conclude

starting at position n, the chance of safety is n/L.

There is a very similar argument to find the average number of steps in the whole walk. Write $T(k)$ for this average starting from point k. Starting at either endpoint, the walk is already over, so $T(0)$ and $T(L)$ are both zero. Corresponding to $(*)$, looking at the result of one step from the intermediate position k leads to

$$T(k) = 1 + \frac{1}{2}T(k + 1) + \frac{1}{2}T(k-1), \qquad (**)$$

another difference equation whose solution is $T(k) = k \times (L - k)$. (It is easy to check that this is correct when $k = 0$ or $k = L$, and that $(**)$ is also true.) So

the average number of steps in the whole walk, starting from n, is n(L − n).

average can only be 1000 if there is a real prospect of a very long run for your money. This example should be fair warning that the *average* length of a game may not be a good guide to its *typical* length.

Even though these calculations assume an unrealistic fair game, they can give indications of what may happen in a genuine casino. In a fair game, the chance of increasing your capital by a factor of ten before you are ruined is one-tenth, however you arrange your bets. So where there is a house advantage, the chance must be *less than* one-tenth, whatever you do.

Take the cautious gambler who places unit bets on Red, hoping to reach some target before he is ruined. This corresponds to a drunkard with a small list to the left, hoping to reach home first despite the bias towards the canal. The boxed argument above can be altered by replacing the equal chances of moving left or right by amounts that reflect this bias. The two equations that correspond to (*) and (**) can be solved in the same way, and the final answers are shown in the next box.

A real casino

Earlier, it was noted that on the bets on Red, it could be useful to modify the zero rule so that your chances of winning or losing your entire stake became 73/148 and 75/148. In some countries, zero is simply a losing bet, so that these chances become 18/37 and 19/37.

To give an analysis that can be used in any casino, write p as the chance you win the bet, and so $1 - p = q$ is the chance you lose. The key quantities are the *ratio* and the *difference* of these two values, so write $x = q/p$ and $y = q - p$. Since bets always favour the house, q exceeds p, hence x is bigger than 1, and y is bigger than zero. Using the same notation as in the previous box, so that (*) and (**) are changed in an obvious fashion, the two answers can be worked out as

$$\text{the chance of 'safety' is } \frac{x^n - 1}{x^L - 1}$$

and

$$\text{the average number of spins is } \frac{n}{y} - \frac{L}{y} \times \left(\frac{x^n - 1}{x^L - 1} \right).$$

Trying some particular numbers is the best way to make sense of these expressions. For UK casinos, take $p = 73/148$ and $q = 75/148$, so that $x = 75/73$ and $y = 1/74$. If our football fanatic Paul had been ill advised, and persuaded to bet one unit at a time on Red in order to turn his initial capital into £216, Table 10.1 shows how little chance he would have of success. The Table uses $L = 216$, with a range of values for his initial capital.

Table 10.1 The probabilities of reaching the target of 216 units, and the average duration of play, when unit bets on Red are made from various amounts of initial capital.

Initial capital	54	90	108	144	162	180	198
Prob. of success (%)	1	3	5	14	23	38	61
Average duration (spins)	3800	6200	7200	8400	8300	7300	4850

This shows the result of timid play, even on the bets on Red where the player's chances are highest. Paul would reduce his chance of doubling his £108 from over 49% (bold play) down to 5%. The consolation is that he makes the casino work for its money, as he ekes out his capital for an exhausting 7200 spins, on average. Timid play keeps you at the roulette table a long time, but the house edge siphons off your capital as surely as the tides slow the earth's rotation—and much faster.

Using physics

After the roulette wheel has begun rotation, and the croupier has flicked the ivory ball in the opposite direction, the laws of physics determine in which slot it will end. Some gamblers have attempted to build on this observation by using hidden computers to calculate the velocities of the ball and the wheel, and to seek to estimate where the ball will meet the wheel. Their intention is to divide the wheel into segments, use their calculations to predict the most likely segment, and cover all the corresponding numbers with single bets. For example, if the computer selected '10' as the single most likely destination, equal amounts might be placed on (say) 24, 5, 10, 23, and 8 (see Figure 10.1). Such a bet would be 'Ten and four neighbours'.

Casinos ban players from bringing computers into the salon, and the call of 'No more bets' is made before the ball has slowed too much. The house edge is so small that it could be overturned if punters could accurately pick out some ball destinations as more or less likely than others. Suppose, for example, that some segment of four numbers, {0, 32, 15, 19}, has only one-half of its expected chance. That alone is enough for a collection of single bets on the other 33 numbers to give a punter a 3.1% edge.

Do not underestimate the complexity of the physics calculations that are necessary. Thomas Bass's *The Newtonian Casino* spells out an attempted route to success. You would need a sequence of readings to find the speeds of rotation of the wheel, and of the ball. You must take account of the small baffles that may deflect the ball as it slips down towards the rotating inner cylinder. Chaos theory shows that even tiny changes in the initial velocities of the wheel and ball can have a large effect on the ultimate destination. But from a punter's standpoint, a precise answer is not necessary. A broad indication of where is more likely, and where is less likely, can be enough to turn the game in his favour. The only other requirements are enough capital to stand an unlucky streak, and lots of patience.

But if you wish to continue to be welcome in the casino, you must obey the house rules, and make no attempt to conceal your actions. The college kids in *The Newtonian Casino* hid their computers in special shoes, tapping buttons with their toes so that the velocities of the wheel and ball could be found. They had some success but did not make a fortune, and their methods were eventually banned. But they had fun.

From the other side

Commercial casinos are reckless gamblers in comparison with the game played in Aden in 1930, described in Evelyn Waugh's *Remote People*. The banker dealt five cards face down, gamblers staked one anna on some card. When all the cards had one backer, the banker revealed the winner, and paid out *even money*! Every game had a guaranteed house profit of 60% of the stakes!

The ultimate dream of a casino manager is to see several players spread their bets around the board, collectively staking equal amounts on all the numbers. The house simply extracts 2.7% of total stakes, and redistributes the remainder back to the players, hoping they will repeat their actions. His nightmare is one gambler placing high stakes on few numbers, which might wipe out a month's profits over a few lucky spins. Averages rule, in the long run, but a casino needs sufficient capital as a buffer against short-term setbacks. The Gaming Board insists that casinos place in a Gaming Reserve a substantial sum, related to the maximum permitted bet. This bond is a safeguard for emergency use only, as casinos are expected to pay out to punters on the spot.

Setting a house maximum caps the loss on any spin. Gamblers have been known to collude, each placing the maximum on the same number, but house rules explicitly prohibit such actions. Just as an insurance company reserves the right to withhold payment when its clients conceal relevant information, so a casino expects gamblers to act honestly. No collusion, and certainly no sleight of hand to 'adjust' a bet after the winning number has been chosen.

A careless casino, that failed to keep its roulette wheels clean and free-spinning on level surfaces, would risk creating an exploitable bias in the outcomes. These wheels are machined to tight specifications, are regularly oiled and can be moved around the room to help make the outputs on any one table completely unpredictable. Casinos do not routinely record the sequences of winning numbers, so if you have any belief that such information will help you select winning bets, you must watch the wheel yourself, or buy an unofficial record.

Croupiers swiftly become expert at arithmetic. When the winning number is known, all losing bets are gathered and sent down a chute that automatically sorts the chips by colour. Suppose 15 is the winning number, and one gambler has placed unit bets on 15 itself, two relevant corner bets and the winning street; the croupier knows that means a payout of 62 chips, in addition to the stakes, but sums 35, 8, 8, and 11 to check his instinct. Just as expert darts players have no need to pause to identify the closing double, so croupiers meet the same arithmetic so often that the winning chips are counted instantaneously.

Although the bets on Red offer better value for money, they are not particularly popular. On purely arithmetical grounds, the zero rule makes one bet of 18 units on Red superior to 18 separate straight up unit bets. But contrast the psychology: if you win with a stake of 18 on Red, you keep that stack of chips—perhaps leaving it there for the next spin—and the house pushes over a stack of equal size. But when one of your straight up bets is successful, a single stack of 35 chips comes your way (while 17 losing chips go unnoticed down the chute).

Were casinos permitted more freedom to advertise, they could make a telling contrast. A gambler might begin to play in a casino that will accept bets of £1, intending to move on to more salubrious ones prepared to pay out up to £2 million. By placing the accumulated stakes on the next number, four winning straight up bets would turn £1 into £1.7 million. The odds are slim, about 1 in 1.9 million—but that is *seven times* as favourable as the odds against winning a share of the National Lottery jackpot, averaging £2 million!

Fyodor Dostoyevsky

Dostoyevsky was a compulsive gambler at the roulette tables, especially in the German spa towns from 1863 to 1871. He wrote *The Gambler* in 1866, drawing deeply on personal experience. His letters over that period assert that the secret of winning is not to get excited, but simply to keep one's head. 'That's it, and by following it it is impossible to lose, and you're certain to win.' He repeated this sentiment several times; in May 1867, writing from Homburg, his conviction is total. 'If one is prudent, that is, if one is as though made of marble, cold and inhumanly cautious, then definitely without any doubt one can win as much as one wants. But you need to play for a long time, for many days, contenting yourself with little if luck is not with you.' 'If you play coolly, calmly and with calculation, there is no chance of losing.' 'Anyone who gambles without calculation, relying on chance, is a madman.' 'By gambling a little at a time, every day, there is no chance of not winning. That's certain, certain.'

Holding these beliefs, he made himself destitute time after time, but was able to rationalize his losses, claiming that over-excitement led to departures from his winning system. Many of these letters were pleas for more money. In one visit to Wiesbaden, he played roulette with the aim of winning one thousand francs to see him through the next three months: unsurprisingly, he lost his capital within five days. By the time of his final gambling bout, in April 1871, he was no longer sure of the inevitability of winning if he kept to a system. It seems that he realized at last that if he set out to win definite amounts to pay his debts, or provide for the future, ruin was far more likely. But how much he stopped of his own accord, and how much was due to the closing of the German casinos by official decree and his early return to Petersburg, is a matter for supposition!

The Gambler is generally sound on the mechanics of roulette, but contains a slip in Chapter 14. Dostoyevsky's *alter ego*, Alexis, describes placing a winning bet on the dozen middle numbers, but he makes the curious claim that this bet 'pays three to one, but the chances are two to one against'. The casino that pays three to one on bets that win 12 times in 37 is yet to be built! Alexis goes on to state that this winning bet turned 80 *friedrichs d'or* into 200, which is consistent with a payout of three to *two*: but why would any gambler accept such odds, when all the other bets were paid at the odds shown in Figure 10.2?

Casino stud poker

Unlike the poker seen in B Westerns, or the official world championships, this version has no element of bluff. It is a sanitized version of a game available on Caribbean cruise ships where a progressive jackpot gave the promise of high rewards, and was introduced quite recently into UK casinos. It accounts for about 6% of turnover. There is one place where you may exercise some skill—whether or not to raise your initial bet—but otherwise the game is pure chance. To remind you, a poker hand consists of five cards drawn from an ordinary deck, and so (Appendix I) there are $^{52}C_5 = 2\,598\,960$ possible hands. These hands are classified as listed below. The number after each heading notes how many hands fall into that category (see Appendix I for justification), and the odds are explained later.

Straight Flush (40): consecutive cards, all in the same suit. Odds 50:1.

Four of a Kind (624): contains four cards of the same rank. Odds 20:1.

Full House (3744): three cards of one rank, two of another. Odds 8:1.

Flush (5108): same suit, not all consecutive. Odds 6:1.

Straight (10 200): consecutive cards, not all in the same suit. Odds 4:1.

Three of a Kind (54 912): three cards of one rank, others not a pair. Odds 3:1.

Two Pairs (123 552): two separate pairs of cards of the same ranks. Odds 2:1.

One Pair (1 098 240): one pair of the same rank, three loose cards. Evens.

AK (167 280): Ace, King, three others; no pair, flush or straight. Evens.

Worse than AK (1 135 260): obvious. Odds not relevant.

This list is in rank order, best to worst. Over 92% of hands are no better than One Pair. Within any category, there is an obvious ordering (Aces are high): thus a Full House of {8, 8, 8, J, J} beats {6, 6, 6, K, K}. Very occasionally, two hands are exactly equal—perhaps both players hold some {A, A, 7, 6, 2}. This is termed a 'stand off'.

The dealer holds the casino's hand, and plays it against each player separately, so it is enough to imagine you are the dealer's sole opponent. Before you see any cards, you select the size of your initial bet, the ante. The hands are dealt face down, except that the dealer's last card is exposed. You now look at your hand, and must decide whether to raise your bet. If you think your hand is not worth a raise, you fold and concede your ante to the dealer, without either of you exposing your cards.

If you wish to continue, you raise by making an extra bet of exactly twice your ante. The betting is now complete, and you have two separate stakes, an ante of one unit and a raise of two units. The dealer exposes her hand. If it is 'Worse than AK', you win, no matter how good or bad your own holding. But in this case, the dealer only pays out even money on your ante, and your raise is simply returned to you as though it had never been made. This will happen quite frequently, since nearly 44% of all possible hands are worse than AK. A player purring over his Full House, supremely confident of winning, may watch the dealer expose a miserable {J, 10, 8, 5, 3}, and have to accept a payout only on his ante.

When you have raised and the dealer holds AK or better—a qualifying hand—both your stakes are in play. If the dealer's hand beats yours, you lose both your bets; with a stand off, both stakes are returned to you; if you win, you are paid even money on your ante, and a bonus on your raise, at the odds listed above.

Check that you have mastered the rules by assessing what happens when you place one unit as ante, and:

A: You hold Two Pairs, and raise. The dealer has {A, Q, 10, 9, 5, not a Flush}.

B: You hold a Flush, and raise. She holds One Pair.

C: You hold a Pair of Twos, and fold. The dealer has the same as in A.

D: You hold Three of a Kind, and raise. She has a Straight.

The answers are:

A: Her hand does not qualify. Your profit is one unit on your ante, your raise is returned unused.

B: She qualifies, and you win. Your total profit is 13 units, made up of one unit on your ante, and a bonus of 12 units (two units staked at odds of 6:1) on your raise.

C: You lose your ante of one unit. You never discover that you would have won.

D: She wins. You lose your total stake of three units.

You never lose more than three units, even if the dealer has a Straight Flush. The most you can win is 101 units, when you win holding a Straight Flush, and the dealer has a qualifying hand. Do not hold your breath: if you play one hand a minute for one hour a day, 250 days of the year, your average wait for a Straight Flush is over four years. And when you hold one, the dealer qualifies only 56% of the time! Your average wait is nearly eight years to collect the 101 units.

You can take it for granted that the game favours the house. When you play this game, there are four questions to answer:

- On which hands should I raise?
- How can I use the dealer's up card to my advantage?
- How big is the house edge?
- What prospects do I have of·a really substantial win?

It would be a lot easier to answer these questions if the dealer's hand were dealt independently from a different pack! If that were the case, our own holding would not affect the possibilities for the dealer in any way. But reality is different. Our hand of five cards leaves 47 for her, so she holds one of the $^{47}C_5 = 1\,533\,939$ equally likely hands that could be dealt. Although this total number of hands she can hold is fixed, *what* those hands can be is affected by what we hold. Annoyingly, our own hand makes only a small difference to the dealer's possibilities much of the time, but on some hands the effect is substantial.

For example, although four Aces and a King will beat four Queens and a Jack, I would strongly prefer to be dealt the second holding. With either hand, I am overwhelming favourite to win, but in order that I win 41 units rather than just one unit, the dealer must hold a qualifying hand. If I hold all four Aces, she cannot have an AK hand, and more of her hands that are worse than One Pair will not qualify. Counting tells us that whenever we hold all four Aces, or all four Kings, the chance she qualifies is under 40%; with Four of a Kind at any lower rank, this chance increases to about 47%. For the same reason, although I would never turn down an AKQJ10 Straight Flush, I would graciously swop it for the corresponding QJ1098. To win a substantial amount, I have to hold Four of a Kind or a Straight Flush, which only occur one hand in 4000; when I do so, my chance of defeat is slim, but I do want the dealer to hold a qualifying hand.

To decide whether to raise or to drop, we should look at the consequences of the two alternatives, and select the one with the better average outcome. (This is the same criterion that we used in Chapter 5, when deciding whether to accept a double in backgammon.) If we drop, we simply forfeit one unit, so we should raise whenever, on average, the result is better than this. To find this average outcome, we embark on an exercise in counting: we want to know how many dealer hands are:

(1) non-qualifying

(2) qualifying, but worse than ours

(3) qualifying, and a stand off

(4) qualifying, and better than ours.

Counting (3) will usually be straightforward. The total of all four categories is fixed at $^{47}C_5$, or at $^{46}C_4$ if we pay attention to the dealer's up card, so the real work is counting (1) and (2). This may well be very intricate; changing one card turns a mediocre J10873 hand into a strong J10987. If the cards are labelled from 1 to 52, i.e. from the Two of Clubs to the Ace of Spades, and all possible hands are listed using a simple algorithm based on that labelling, the order in which they arise bears little relation to their strengths as poker hands.

Consider the two AK hands:

Hand A: Spade Ace, Heart King, Diamond Jack, Club Eight, Spade Two

Hand B: Heart Ace, Spade King, Heart Jack, Spade Eight, Heart Two.

If we hold one of these hands and the dealer has the other, the outcome is a stand off: in that sense, the hands are equal, as neither beats the other. But Hand A is better! If we hold A, rather than B, it turns out that the dealer can hold more non-qualifying hands, more that qualify and are worse than ours, more stand offs, and fewer hands that beat ours. A complete justification of that statement requires an exhaustive enumeration of what the dealer can hold in the two cases, but the heart of the matter comes down to counting the numbers of Flushes possible. With either hand, the numbers of dealer hands that are One Pair, Two Pairs, Three or Four of a Kind, or Full House are exactly the same. But Hand B leaves the rest of the pack richer in Flushes than Hand A—and any Flush swamps either hand. The Flush deficiency in A is overwhelmingly distributed among dealer hands that do not qualify, and are thus worse than Hand A. As George Orwell nearly put it, hands A and B are equal, but A is more equal than B. The box (next page) gives the details of the Flush calculation.

This example is a reminder that exact calculation on this game will be complex. I emphasize 'exact': raising on B leads to an outcome that, on average, is only 1/1000 of a unit worse than raising on A. But we cannot know this without doing the sums—and the blood rises to the challenge of deciding exactly which hands are worth a raise and which are not.

In this game, the dealer never bluffs. She must fold and concede one unit if she holds worse than AK, and must play otherwise. It seems plausible that if some particular hand is worth a raise, then any hand that beats it is also worth a raise; and if we should drop on some hand, we should also

---

Hand B leaves more Flushes than Hand A

Hand B leaves 10 Hearts, 11 Spades, 13 Clubs and 13 Diamonds, so the number of Flushes is $^{10}C_5 + {}^{11}C_5 + {}^{13}C_5 + {}^{13}C_5 = 3288$.

Hand A leaves 11 Spades and 12 in each of the other suits, so there are $^{11}C_5 + 3 \times ({}^{12}C_5) = 2838$ Flushes, 450 fewer.

Careful inspection shows that B leaves 27 Straight Flushes (four in Hearts, three in Spades, ten in each other suit), while A leaves 26 Straight Flushes (count them!). That means B leaves one extra Straight Flush, and 449 extra ordinary Flushes for the dealer. As it happens, A leaves one more Straight than B, but otherwise B has 360 fewer non-qualifying hands, 43 fewer AK hands that are worse, two fewer stand offs, and 44 fewer AK hands that beat A or B.

---

drop on any hand that would lose to it. That leads to the notion of a *cut-off hand*: any hand that beats this is worth a raise, any hand that loses to it should be dropped. Despite the interaction between our holding and the dealer's possibilities, that notion turns out to be correct, if we ignore the dealer's up card. However, when we do pay attention to this up card, matters are much more messy: on some hands that beat this cut-off hand, certain up cards will persuade us to drop, while on some hands worse that lose to it, certain up cards favour a raise. We will proceed by identifying the cut-off hand, and then outlining how to modify the decision according to the up card.

This is a nice exercise in informed guesswork—circumlocution for trial and error. To begin with, ignore the up card, and the effect of our hand on what the dealer can hold, and disregard stand offs (which are rare). We want to find some hand that is clearly worse than the cut-off hand, another one that is better, and then home in on the target with a pincer movement. Exact analysis will come in later, but there is no point in working to five decimal places when all we seek is a clear indication of whether the loss is rather more than one unit, or definitely less.

Suppose all possible poker hands are ranked in order. If our hand is in the top half, we expect to be favoured, while if it is in the bottom half, the dealer seems more likely to win. Make a stab at the cut-off hand as being a bit below the middle-ranking hand. Since 44% of all poker hands are worse than AK, we explore the consequences of raising with the weakest possible AK hand, some AK432.

Ignoring the effect of our holding on the dealer's possibilities, she will not qualify 44% of the time, conceding one unit to us. But the 56% of the time she does qualify, we lose three units as her hand inevitably beats ours. Our average outcome is

$$(1) \times (0.44) + (-3) \times (0.56) = -1.24$$

which is 24% worse than quietly folding and losing one unit. It looks as though we need a better hand than AK432 to raise.

About 50% of hands are One Pair or better. See what happens if we raise with the best AK hand, AKQJ9. Here we win one unit if the dealer does not qualify, or three units if she has some AK hand, but lose three units if she holds at least a Pair. With the same crude assumptions, but working to one more decimal place, the average outcome is

$$(1) \times (0.437) + (3) \times (0.064) + (-3) \times (0.499) = -0.868$$

which is about 13% better than dropping and losing one unit. The best AK hand looks worth a raise.

If our assumptions are not so crude as to mislead, we have already made a decision about the 93.6% of all hands that are not AK. Drop if worse than AK, raise if better than AK. Notice that the estimated 24% loss through raising with AK432 is about twice as big as the 13% profit through raising on AKQJ9, which suggests the cut-off hand is about two-thirds of the way up the ranking list of AK hands—some AKJxy hand. It is not worth pressing much further, as our small card holding will become crucial, but we have made substantial progress, with little real effort.

When we raise with either an AK or a One Pair hand, we win or lose exactly three units whenever the dealer qualifies. So suppose our hand is no better than One Pair. Given that hand, let N be the total number of possible hands for the dealer, and consider what she might hold. Write:

x = Number of dealer hands that are worse than AK

y = Number of AK or One Pair dealer hands that are worse than our hand

z = Number of dealer hands that are a stand off

leaving N-x-y-z dealer hands that beat our hand. Since all dealer hands are equally likely, the *average* amount we win if we raise is found in the standard fashion (Appendix III) by weighting the outcomes that we win 1, 3, 0, or −3 with their respective probabilities. The arithmetic leads to an average profit of

$$\frac{4x + 6y + 3z}{N} - 3.$$

We ought to raise whenever this amount exceeds -1, the amount we get if we drop. This points to the decision criterion: raise provided

$$4x + 6y + 3z > 2N. \qquad (*)$$

Recall Hands A and B described earlier. By careful counting, $(*)$ is *just* satisfied for A, but fails for B. And examining all other possible AKJ82 hands, we reach a neat conclusion: *raise that hand if all four suits are present, drop if only two or three are present.*

Moreover, every hand that loses to AKJ82 should be dropped, every hand that beats it is worth a raise—on average.

For any hand, what happens overall is the average outcome over the 47 possible dealer up cards. Some of these up cards will encourage us to raise, others will deter us. The fine details are complex, but are described by Peter Griffin and John Gwynn, who examined the interactive combinations of the player's hand with the dealer's up card and the ways of choosing the rest of the dealer's hand—nearly 20 million million combinations. The criterion remains as $(*)$, where now, since we are using the up card, N is 163 185, the number of ways of selecting the last four cards for the dealer from the 46 cards left.

In broad terms, the factors (in order) that influence us to raise on marginal hands are:

- the up card is the same rank as a card we hold;
- the up card is low rather than high;
- we have cards in the same suit as the up card.

Why are these considerations important? For the first one, suppose the up card is a Seven. If we also hold a Seven, the dealer's chance of making a Pair is reduced in two ways: *our* holding makes her less likely to make a Pair of Sevens, and *her* Seven cuts down her chance of making a Pair of a different rank. And a Pair is the most likely way her hand will beat ours. The second consideration is obvious (but often far less important than the first), and if we have cards in the same suit as her up card, we cut down her chance of holding a Flush.

You will understand why the details of the counting are omitted. If you want the complete best strategy, you should consult the account by Griffin and Gwynn. For any hand and up card, $(*)$ is the criterion, but noting which ranks and how many different suits are present turns out to be not always sufficient: sometimes, you have also to name every card in the hand (apart from total suit permutations). Using such detail, tiny improvements to the strategy described below can be found, but you would have to play for large stakes and a long time to notice them. Points (1), (2), and (3) below are always correct; points (4) and (5) are a good approximation to best.

For a good easy-to-remember strategy, you should:

(1) Always raise when you hold AKQJx or better.

(2) Always drop if your hand is worse than AK.

(3) With any AKJxy hand, look at the up card: raise if has the same rank as any card you hold (15 chances), drop otherwise (32 chances).

(4) An AKQxy hand is better than the corresponding AKJxy hand. As well as when you hold a card with the same rank as the up card, you should also raise when the up card is low enough. For example, with AKQ94, only fold when the up card is Ten or Jack; with AKQ54, only fold when the up card is a Six to a Jack.

(5) With AK10xy or a worse AK hand, drop when the up card is Ace, King or a rank you do not hold, and sometimes on other occasions. This means there are at most 9 of 47 possible up cards where you will raise. Indeed, as an example, holding AK932 you should raise only when the up card is Two or Three, just six cards.

Griffin and Gwynn calculate that the decrease in average loss by paying attention to the up card, rather than ignoring it, is under 1/1000 units overall, so you might well regard this five-point elaboration with detached amusement. Nevertheless, now that the calculations have been done, you have the comfort of *knowing* how little extra information the up card brings!

If you follow this strategy, you will drop on nearly 48% of all hands, and raise otherwise. That means your average stake will be

$$(1) \times (0.478) + (3) \times (0.522) = 2.044.$$

Griffin and Gwynn used computer enumeration to calculate that the average loss using the strategy described above is about 0.05 units. So the percentage loss on your average stake is about $(0.05)/(2.044)$, say 2.5%— very comparable with roulette.

Generally, the higher up the ranking your hand fits, the better you can expect to do, on average. But this is not true at every stage. We noted earlier that four Queens would lead to a big win more often than four Aces, but we do not have to look to such spectacular hands to find anomalies. One question of mild interest is to find which hands lead to an average outcome that puts us ahead. Can we find a cut-off hand below which we lose, on average, but above which we expect to gain? The answer is no.

Detailed calculation shows that if you raise with the One Pair hand 66987, you will be ahead, on average. Although the hand 661052 beats 66987, it leads to an average *loss*. Even without going into detail, it is not difficult to appreciate why 66987 does better than the other hand. Either hand leaves just as many One Pair hands for the dealer: if you hold 66987, a good number of the dealer's One Pair hands are when his Pair is 2, 3, 4, or 5, all inferior to your hand. But when you have a 661052, your own holding cuts down his chance of a Pair of Fives, or a Pair of Twos, by enough to make a difference. There is no clean break between hands that make a loss and those that make a profit.

Whoever invented this game showed a fine imagination. Novices need not worry about whether any players are bluffing, and they are not at a disadvantage because of some opponent's sizeable capital. But the inventor made sure that, with all its odds and intricacies, this game was definitely in favour of the house.

Baccarat chemin de fer

This is a simple game, and quite popular, accounting for about one sixth of the drop in casinos. The player and the bank each receive two cards, face down. Each card takes its face value, except that court cards count as zero. When computing the total of a hand, multiples of ten are disregarded. Thus a Nine and a Six, or a Queen and a Five, or an Ace and a Four all lead to a total of five. The winner is the one with the higher total: equal scores are a stand off.

The player may stand, or draw a third card, but under severe restrictions. He must draw to a total of four or less, he must stand on six or more, but has a free choice with a total of five. The reason for this straitjacket is to protect the interests of any other players who are betting on his hand. Any card taken is dealt face up.

To end the story first, and describe the plot later, I will state now that when the player holds a total of five, he should draw or stand at random, with respective probabilities (9/11, 2/11). That is best play, and leads to an advantage of 1.28% to the bank, rather less than in roulette or casino stud poker.

One way he might achieve this strategy is to carry a supplementary pack of cards from which the Kings and Queens have been removed, leaving 44 cards, in 11 ranks. Before the baccarat cards are dealt, he privately allocates two ranks to stand—Four and Ace, say—and cuts the deck at random. This card determines his decision, should his total be five. Since he alone knows which ranks are allocated to stand, he is indifferent to prying eyes. Next hand, he can stay with Four and Ace, or change his mind. All he seeks is a reliable way to obtain a chance of exactly 2/11. But he should do this before each hand; if he only went through it after he knew his total were five, his holding would be an open book.

Now for the plot. After the player has exercised his choice to draw or stand, the banker has the same option. To help her play well, the casino will give her a list of instructions as to the action to take, according to her total and the result of the player's action. To see why such a list is necessary, consider the banker's dilemma without it. Her own total can be one of eight alternatives, zero to seven (two-card totals of eight or nine are declared at once, without any draw). There are 11 possibilities for the player: either he did not draw, or he did and shows a card with a score from zero to nine. The banker faces $8 \times 11 = 88$ different circumstances. And that is just the beginning: a single 'strategy' for the banker is a list of the 88 alternative choices, D(raw) or S(tand), according to what her hand is and what action the player took. This gives her 2^{88} different strategies to choose from, a number with 27 decimal digits!

Thank goodness for game theory. Recall, from Chapter 6, the idea of a *dominated* strategy, one that can be eliminated because some alternative strategy never does worse and sometimes does better. Here, for example, whenever the banker has been dealt a pair of Fives giving a total of zero, it would be sheer lunacy not to draw. Drawing can only improve the hand. It turns out that the overwhelming majority of conceivable banker strategies can be struck out using this idea, and there are just four places where it is not clear what the dealer should do. They are:

E: Her total is six, and the player did not draw.

F: Her total is five, and the player drew a Four.

G: Her total is four, the player drew an Ace.

H: Her total is three, the player drew a Nine.

She has two choices for each of these, which reduces the number of strategies to $2^4 = 16$. Having got as far as this, most of these 16 can now also be thrown out, as they are dominated by one of the others, or a mixture of two others, leaving just five. Pause to admire game theory: out of that vast original number of possible strategies, dominance leaves five worth serious consideration.

The second useful notion from game theory is that when one player has only two choices (as here), there is a best play by the opponent that uses only two of her choices. For this game, two such choices for the banker are SDSD and DDSD at the four problem positions listed above. These give the same play to alternatives F, G, and H, and so the banker can achieve best play by varying her selection only at E, i.e. when she holds a total of six and the player did not draw.

In summary, at 87 of the 88 possible decisions she can face, the banker can glance at a printed card provided by the management, and draw or stand to order. In the 88th position, E, she should randomly draw or stand at frequencies to be determined. The arithmetic comes from the table of payoffs, which has the form:

Player/Banker	Draws when E happens	Stands when E happens
Draws on Five	–4121	–3705
Stands on Five	–2692	–4564

(The actual payoff to the player is the entry in the table, multiplied by 16/4826809; we saw in Chapter 6 that we can get rid of fractions in this way.)

Each entry in the table is the result of a large averaging process. That in the NW corner corresponds to all possible deals, correctly weighted, with the player always drawing to five, and the banker always drawing on a total of six when the player did not draw. You will understand why I am suppressing the details. This game is normally played with six packs of cards, and the calculations leading to these results make the assumption that the cards held make no significant difference to the chances of drawing from the rest of the pack.

All the above was worked out by John Kemeny and Laurie Snell and published in 1957. (Kemeny and Snell achieved even greater fame later through their work in the development of the computer language, BASIC, at Dartmouth College.) The best strategies for the player were given earlier, and the banker should draw or stand with respective probabilities in the ratio 859:1429. This happens to be very close to the frequencies (3/8, 5/8). Practical ways to achieve this particular mixed strategy were given in the earlier chapter. Frank Downton and Roger Holder, in their 1972 analysis of casino games, suggested an even easier way for the banker to achieve her mixed strategy, by taking account of the card combination that led to her total of six.

If you have been very assiduous, or are independently knowledgeable about game theory, you will remember that the best play by one side in a game where both players have just two choices automatically protects the opponent from any folly they commit. So if the banker acts in the way identified, it does not matter what the player does when his total is five; and if the player acts optimally, the banker can act arbitrarily when circumstance E arises. But how much could a banker increase her advantage if the player failed to use the correct mixed strategy? Recall her normal advantage is just 1.28%. However, if the player with a total of five:

(1) always stands, then the banker should stand with E—advantage 1.54%

(2) always draws, then the banker should draw with E—advantage 1.37%

(3) uses a 50:50 mixed strategy, the banker can also achieve 1.37%.

Small changes, agreed, but why give *any* money away through ignorance or wilful disregard? Just as you might fail to use your best strategy, there is always the hope that the banker may not act optimally when she has this choice we have labelled as E. Perhaps she consistently stands, or she consistently draws, at that time. Turn back to the table of payoffs, and look down each column. Plainly, if she always draws, then you should always stand on five; and if she always stands, then you should always draw on five. Your average payoff remains negative, but not *quite* as negative. Every little helps.

Other casino games

About one-sixth of the drop in casinos is used to play some form of Blackjack, and about 1% of money staked is on the dice game, Craps, mentioned in Chapter 5. Blackjack has received enormous attention from mathematical analysts armed with computers, ever since Edward Thorp's *Beat the Dealer* was published in 1962. The consensus is that, playing optimally and invoking card-counting strategies, a player *can* engineer a small advantage over the bank. Martin Millman, writing in 1983, estimated that with four decks of cards the player could gain 1.35%, with six decks that reduces to 0.91%. But the details of the optimal play are very intricate: they consist in stating whether you stand, draw, split, or double down on every possible combination of your own holding and the dealer's up card—a large matrix of instructions. And even though the player can achieve this advantage, it is small enough to leave him a considerable risk of accumulating substantial losses in the medium term.

If you have a serious intention of learning the best play at blackjack, you should consult a specialized book that both describes the play, and justifies it so that you understand why you are making certain plays. I admire the attitude of A. P. Herbert, the eccentric wit who was an Independent MP for several years. In his election address, under the heading 'Agriculture', he described his policy simply: 'I know nothing about agriculture.' My aim in this book is to help you *understand* probability, and since I cannot explain the details of best play at blackjack in a way that enhances this aim, take it that 'I know (almost) nothing about blackjack'.

But details aside, the bones of a successful strategy, and their rationale, can be described. You need a method of judging whether the cards left in the shoe, that you or the dealer might draw, will tend to favour you or her. When you are favoured, you bet your maximum, when she is favoured you either leave the table, or bet your minimum. The ability to vary the stake is a central plank of a winning strategy, and you either bet low or high; only wimps bet moderately.

Casino rules instruct a dealer to stand with a total of 17 or more, to draw with 16 or under. You hope the dealer will bust. The circumstances most favourable to this are when she holds just under 17, and the remaining cards are rich in Tens, but relatively thin in Fours, Fives, and Sixes. Millman's advice is to watch all the cards dealt from the beginning, after a new shuffle, and count as follows: starting with a total of zero, add one point whenever a Two, Three, Four, Five, or Six is dealt, ignore any Seven, Eight, or Nine and subtract one point when any Ace, or card counting as a Ten, is dealt. The bigger this running total, the more favourable your position. (This plainly makes sense: there is a big total only if a lot of low cards have already gone, but not as many high cards, and this will help the dealer go bust.) To take account of how many cards are left in the shoe, divide this running total by the number of 'packs' still left in the shoe. The result of this arithmetic is some number, T. Early in the game, the count will have to be quite large for T to be big, as there are plenty of cards left.

Do not bet when T is negative, bet the minimum when T is less than two, bet the maximum when T is at least two. Exactly where these numbers come from, and what you should do holding a pair of Nines when the dealer's up card is a Seven, and all your other decisions, you must look elsewhere. I know nothing about blackjack.

Test yourself quiz

10.1 Suppose you have £20, and wish to maximize the chance you can turn it into £100 at the casino. Describe exactly what 'bold play' will consist of, if you restrict yourself to bets on Red. What will your chance of success be?

Compare this strategy with that of betting £16 on a block of six numbers (which pays odds of 5–1), and if this loses, placing your last £4 on zero.

Could you do better than either of these, if your sole intention is to have the best chance to hit £100?

10.2 You take £100 into the casino, and make a series of £1 bets on your pair of lucky numbers, {21, 24}, for about one hour. The number of spins will be about 90. What happens, on average? Compare your chances of being ahead after 89 and after 91 spins.

10.3 In the drunkard's walk, suppose the street is 100 steps long and the pub is 20 steps from home. What is the chance he reaches safety? On average, how many steps does he stagger until his fate is sealed? Do the corresponding calculations for his brother, whose steps are just half as big.

10.4 In casino stud poker, decide what happens if:

A. you hold {Q, Q, K, 10, 4} and raise, the dealer has {Q, Q, J, 10, 9};

B. you hold three Kings and raise, the dealer has a Full House;

C. you hold a Full House and raise, the dealer has three Kings.

CHAPTER 11

Bookies, the Tote, spread betting

Bookmakers

Bookmaking grew up around horse racing, which still accounts for over 70% of the industry's turnover. Wagers have been struck at racing in York, Doncaster, and Chester since the sixteenth century, but the first known professional bookmaker was a Harry Ogden, who worked at Newmarket Heath in 1794. In 1853, betting houses were banned, and for a hundred years only those who attended race meetings could (legally) place cash bets. Otherwise punters had to use credit, and to communicate with their 'turf accountant' by letter, telephone, or telegram. This transparent attempt to confine betting to the wealthier classes was circumvented by the use of runners who collected cash bets in the street. After betting shops were legalized in 1960, illegal bookmaking all but disappeared, and with it went the attempts to bribe the police, protection rackets, and associated gang violence.

So long as betting was legal and not taxed, it was conducted openly and with no fear of unfair competition from dubious or criminal elements. But as betting taxes began to be introduced, and then raised, legitimate operators faced the possibility of losing their trade. The 1978 Royal Commission on Gambling pointed to the danger of higher taxes driving betting activity back to the furtive position before the 1960 Act, where dissatisfied punters had no hope of redress. The current tax, which effectively takes 9% of winnings, is thought to be as high as the legitimate market will bear, although a higher rate operated between 1981 and 1996. To encourage attendance at race meetings, betting on course is free from tax.

Today, the UK has over 8000 betting shops. About 40% of them are small local enterprises, and the rest belong to four major chains, with Ladbrokes and William Hill dominant. Aside from horse racing, greyhounds and soccer are popular subjects for betting. Stung by competition from the National Lottery, the large firms joined together and offered a fixed-odds game based on drawing six numbers from 49. Bookmakers will accept wagers on general elections, the contents of the Budget, or almost any subject that is not in bad taste. Parents have placed bets on the academic or sporting success of their children, but the bookies firmly refused to accept bets on the date that the Gulf War would begin.

You will almost certainly know the meaning of bookmakers' odds expressed in forms such as 5:1, 4:6, 15:8, etc., but I will remind you, for completeness's sake. The first figure is the amount of profit the punter would make, the second is the corresponding stake. If a horse is offered at 4:6, the punter who wants to make a profit of £4 must stake £6. Odds of 4:6 are often described as 6:4 *on*.

A bookie is staking his judgement against that of punters. In a two-horse race, he may believe that the respective chances are 70:30, but he cannot afford to offer the 'true odds' of 3:7 and 7:3. He needs to build in a margin of protection, and so might offer the favoured horse, Alpha, at 1:3, and the other one, Beta, at 2:1. A punter whose judgement coincided exactly with that of the bookie would see neither bet as worth making. But a punter who thought that Alpha's chance exceeded 75% would see the odds of 1:3 as favouring a bet; in the long run, if his judgement is accurate, he will win more than he loses. Another punter, who put Beta's chance at more than one-third, would similarly be happy to bet on Beta.

Whatever the event, the collection of odds offered by bookmakers will have some *margin* or *overround* that you should be aware of. The bigger this margin, the harder it will be to find a favourable bet. To find the margin, take the odds for each horse in the race, and calculate what I will call the 'relevant fraction'. If the odds are 5:1, that fraction is 1/6; if the odds are 10:1, it is 1/11; for a horse quoted at odds of $x{:}y$, the relevant fraction is

$$\frac{y}{x+y}$$

Find this for all the horses in the race, and add these fractions up. The amount by which the sum exceeds 1 is the bookie's margin. In the example above, with odds of 1:3 and 2:1, the relevant fractions are

$$\frac{3}{1+3} = \frac{3}{4} = 0.75$$

and

$$\frac{1}{1+2} = \frac{1}{3} = 0.333\ldots$$

The sum is thus $0.75 + 0.333 = 1.0833..$, a margin of just over 8%. (Strictly speaking, the bookie's expected percentage profit is slightly less than this. He sets the odds in the hope that his liabilities will be roughly the same, whatever the race outcome. If £75 were bet on Alpha, and £33.33 on Beta, the bookie would indeed pay out £100 whichever horse won. He has made a profit of £8.33 on a total stake of £108.33, so his profit margin is really 8.33/108.33, about 7.7%. But we will stick to established terminology, and retain the term 'margin' to describe the figure of 8.33%.) Typical odds for the three outcomes of home win, away win, and draw in soccer are 5:6, 11:4, and 9:4 respectively. The relevant sum is

$$\frac{6}{11} + \frac{4}{15} + \frac{4}{13} = 1.1198\ldots,$$

giving a margin of almost 12%. The Grand National usually has a large field, with considerable uncertainty about the form of a number of the horses, and the margin would be much greater. If a margin is too large, punters will be disinclined to bet, as no horse looks value for money; a small margin encourages bets, but the bookie risks a substantial loss. Bookies compete against the Tote (see below), whose margin on win bets in single races is about 19%.

It is remotely possible that the bookies might set prices that gave a negative margin. Suppose the odds for home win, draw, and away win in a soccer match were 6:5, 11:4, and 11:4. The relevant sum adds up to less than 1, and this always gives the opportunity for the punter to be sure of a profit. On those odds, he could stake £75 on a home win, and £44 on each of the other two results, leading to a £2 profit whatever the match result. Multiply that by 1000, and life looks splendid—except for the betting tax that would wipe out this gain! Only incompetence would allow an individual bookie to make such a mistake, but a punter can sometimes construct a negative margin on an event by taking advantage of different odds from several sources.

The notion of constructing a negative margin is behind attempts to nobble the favourite in a race. Suppose the margin is 12%, and the favourite, Gamma, is priced at 6:4. Then Gamma contributes $4/10 = 40\%$ to the sum of the odds, so if subterfuge could eliminate that horse, there would be a 28% margin in favour of the punter. Careful placing of bets on all the horses except Gamma would then guarantee a profit. The favourite must account for the single largest quantity in the sum of the odds, so that is the right horse to nobble. No-one bothers to nobble the long shot.

Just as a casino manager glows with satisfaction when all the 37 outcomes on a roulette wheel carry equal stakes, so a bookmaker has an ideal: set the prices in such a way that whatever horse wins he is in profit. To appreciate the obstacles to putting that notion into practice, follow the story of the event known throughout racing as 'Dettori Day'.

This was on 28 September 1996, when the jockey Frankie Dettori rode all seven winners at the Ascot race meeting. Gordon Richards had ridden 12 successive winners in 1933, but he had missed some races each day. Willie Carson had ridden six winners out of seven at Newcastle in 1990, but Dettori's feat of going through the whole of a seven-race card was without precedent. The bookmaking industry is reckoned to have lost £40 million through this event, and the immediate reaction was to declare the day an unmitigated disaster. One spokesman said 'The fifth win was expensive, the sixth dismal, and after the seventh it was time to put the lights out.' But within days, the bookies reversed their attitude, and declared Dettori Day 'the best thing that ever happened to racing'. The punters got paid, the media lapped up the story and gave racing free publicity, and the excitement led to renewed interest in the possibility of caning the bookies.

Dettori's feat was so costly for the bookies through a combination of factors. Foremost is the jockey himself. Dettori has a rare ability, and the charisma that captivates punters who know he will ride every horse as though his own mortgage were riding with him. No other jockey would attract so many bets, either single or multiple; no other jockey would attract so many accumulator bets on all seven races. Second was the venue: this was no up-country midweek meeting; this was Ascot, one of the most prestigious courses in the country. Several races carried substantial prizes, with the expected highlight being the third race, the Queen Elizabeth II Stakes. It seemed likely that the outcome of that race would determine the champion trainer for the season. Punters place more bets on outcomes of the more important meetings. Thirdly came television, and timing: the BBC were showing most races live that Saturday afternoon, so punters watched the day unfold, and then joined the Dettori bandwagon as the winners came home.

A professional odds-setter would have expected to see Dettori win at most two, or perhaps three, races all day. Only one of his rides was priced as favourite in the morning betting, and three of them were seen as distinct long shots at odds of 10:1 or more. These odds are set to attract early money, and the off-course punter can place his bet either at the definite odds then stated, or at the starting price (SP). The SP is determined independently, and relates to what is happening on the race course itself. The criterion is the odds that would be available on a substantial bet just before the race started. On that day, the SP was assessed from the prices offered by about half a dozen of the hundred or so individual bookmakers operating on the Ascot course. Because Ascot is just one race course, and betting shops are open throughout the country, the overwhelming proportion of bets are made at these off-course offices, and so have no direct impact on the SP. As will be seen, it was crucial that the off-course bookies were able to use their on-course contacts to drive the SP down. An insurance company would refer to the concept of risk management.

As bets come into the off-course offices, the odds on offer will shift to reflect the weight of money placed. The initial odds reflect the book-maker's judgement of the likelihood of each horse winning, and his assessment of what price will attract bets. If unexpectedly large sums are placed on one horse, its odds will shorten to protect the bookie's exposure, while the odds for other horses will lengthen to attract money elsewhere. These activities are taking place in betting shops in hundreds of locations, so bookies find it difficult to ensure a profit whatever the outcome: a typical position in a 20-horse race could be that the bookies will be in profit should any one of 16 horses win, but will lose on the other four. Sometimes, the bookies will expect to be in profit when the favourite wins.

Dettori's first mount, Wall Street, had begun the day at odds ranging from 5:2 to 3:1, but started as the 2:1 favourite, and won by half a length. No surprise there. His second horse, Diffident, was one of 12 runners and expected to finish well down the field. The odds drifted from the morning price of 10:1 out to an SP of 12:1, but Diffident finished a short head in front of the second horse.

The next race was the 3.20, the main event of the day, but with only seven runners. For one national firm, Coral, most direct money was staked on Bosra Sham, but by now Dettori's horse, Mark of Esteem, was carrying extra money from the accumulator bets on his two previous winners. Across all the horses, the odds on offer gave a bookies' margin of about 12%, and Mark of Esteem's price was 10:3, the joint second favourite. Purely on the direct money on that race, a Dettori win would have left the bookies in profit, but the carryover from the previous races changed the picture. When Mark of Esteem stormed home one and a quarter lengths in front, alarm bells rang. All these wins were being seen by millions of TV viewers, the next race was also to be broadcast live, and Dettori looked invincible.

His mount on the 3.55, Decorated Hero, had been priced at 14:1 in the morning betting, but within minutes of the finish of the previous race, the odds off course were slashed to 8:1. This was expected to be simply a precaution, as there were 26 runners and 14:1 was a reasonable objective price to describe the horse's chances. Decorated Hero romped home, three and a half lengths clear with the SP at 7:1, and the contingent liabilities rocketed. Nothing could change matters for those punters who had been given definite odds in the early betting, but now the off-course bookies began attempts to force down the SPs of Dettori's rides in earnest.

Their strategy was to bet heavily with the on-course bookies. To accomplish this, the major off-course firms telephoned instructions to representatives on course. Correct timing was crucial: substantial sums would be necessary, but if this money came on too early the on-course network would be able to lay off its liabilities on to other shoulders, and the price could drift out again. (An insurance company speaks of 'reinsurance' to spread the risk.) Just as the horses approached the starting stalls, huge bets flooded on Dettori's mount, Fatefully. The horse had begun the day at 10:3, but this strategy drove the price down to an SP of 7:4, the favourite. The activity was very close to being unnecessary, as Dettori's margin over the second horse was returned as a neck, but now the writing was plainly on the wall.

You will appreciate that the bookies' tactics were successful in two ways. Instead of paying out at 10:3, they paid winning punters at 7:4; each £12 bet meant a loss of only £21, and not £40. Furthermore, they themselves had bet heavily on Fatefully, so collected consolation winnings from the on-course bookies.

Dettori had now won five races in succession, and the BBC responded by changing their plans, interrupting the reports on soccer matches to bring the 5.00 race live. What had begun the day as an unnoticed fill-in race with just five runners now became the centre of attention. Of course, even more accumulator money was riding with Dettori, and the previous strategy had to be repeated, in Spades. Two minutes before the start, Lochangel was on offer at 13:8; Coral had bets of over £600K carried over from previous races, and placed over £80K with the on-course bookies to help drive the SP down to 5:4. This act reduced their contingent liability by £225K, and also won the firm over £100K on course. However, there can be few occasions when backers have bet so heavily on a horse, hoping it would finish last! Lochangel won by three-quarters of a length.

The final race was scheduled for 5.30. Hollywood could not have written a more implausible script. A year earlier, Dettori had won the corresponding race, *and on the same horse*, Fujiyama Crest, that he rode again that day. Fujiyama Crest now carried the top weight, and came into the race with a dismal record over the season: he had run five times, had never finished better than third, and had been unplaced in his last two races. There were 18 runners, and Dettori's mount had begun the day at 12:1, a conservative price for a one-off race. But after Dettori's fourth victory, the off-course price for Fujiyama Crest was already down to 7:1, then to 4:1 after race 5, and the absurd price of 6:4 after Lochangel's victory. In the nature of things, on-course bookies have negligible running-on liabilities from previous races, so their prices would reflect their own views of the horse's chances, and current bets. Thousands of pounds showered on Dettori at odds of 7:4, and yet the on-course bookies were willing to *increase* the odds to 2:1, seeking more money. No-one had ever ridden all seven winners before! Surely it was impossible? (This argument is rather like a gambler in a casino watching Red appear ten times in succession, and *as a consequence* betting on Black, on the grounds that runs of 11 Reds never happen.) Fujiyama Crest tore into the lead, and managed to hang on, a neck ahead of the competition, Dettori leaping off the horse in celebration.

Celebrations were muted in the back offices of the bookmakers. Their only consolations were that they had managed to throw so much money on Dettori that the SP had been held to 2:1, and had not slipped to 5:2, and they had their own winnings to collect. But compare that with what might have been! Those punters shrewd enough to have placed a seven-horse accumulator bet at fixed odds early in the day were paid at £235K for a £1 stake: the same bet at SP won £25K. Despite the presence of far better prospects in the fields, the weight of money had made Dettori's horse the favourite in each of the last three races.

Next day's newspapers extolled the rider's feat, while the bookies simply shelled out. A young joiner from Morecambe had staked a total of £64 on a 50p accumulator—all possible combinations of all Dettori's mounts—and was paid £550K. He placed the parallel bet on Dettori's seven rides the next day, Sunday, but lightning did not strike twice. Frankie had to wait until the last race for his first victory. The bookies were already recouping their losses.

Although that day was costly to the bookies, 200 years of racing history had preceded the feat. Just as insurance companies build into their premiums the risks associated with earthquakes, hurricanes and other natural disasters, so the bookmakers drew on their reserves to meet this setback. Despite their best efforts, the off-course bookies were unable to influence

Table 11.1 The 3.20 at Ascot, 28 September 1996.

Horse	Amount bet (£)	On course SP	Bookies' desired price
A	10 500	9:4	4:1
B	21 158	10:3	6:4
C	9080	10:3	9:2
D	6334	11:2	7:1
E	1998	10:1	25:1
F	6430	14:1	7:1
G	2944	25:1	18:1
Total	58 444	Sum 1.119	Sum 1.123

the prices enough to ensure a profitable book, almost irrespective of the race outcome. Further detail from Coral's experience in the main race on Dettori Day will quantify the problem.

A total of £58 444 was directly staked on the seven horses in the amounts shown in the second column of Table 11.1. It does not include money running on from previous races. The SPs, determined from the on-course bookies, are shown in the third column. It is easy to work out the potential profits or losses, according to which horse would win. For example, had horse A won at 9:4, the bookie would pay out (including the stakes)

$$\frac{13}{4} \times £10\,500 = £34\,125.$$

This would have given them a net profit of £58 444 − £34 125 = £24 319, even though this horse was the favourite. The arithmetic for the other horses leads to bookies' profits if C, D, or E had won, but losses if the winner were B, F, or G. The range of outcomes ran from a £38K loss if horse F had won to a £36.5K profit if horse E had been victorious.

Had the bookie been free to fix the odds, subject to approximately the same margin, he might have selected those shown in the final column. With those odds, a profit is certain. The worst that could have happened was victory by G, making £3400, and the best was a profit of £8350 if C were the winner.

The reality was very different. Horse C was Dettori's mount, Mark of Esteem. From Table 11.1 alone, a Dettori win seems to lead to a £19K bookies' profit. But fold in the money from doubles, trebles and accumulator bets from previous races and those still to come, and this 'profit' was more than wiped out. Producing a reasonably balanced book, such as that in the last column of the table, is almost impossible in current betting conditions.

It is salutary to note the aggregate outcome of all seven races that day. Ignoring the complications of multiple bets, and looking only at direct money on each race, wins by 75 of the 93 horses would have meant profits for the bookies, wins by the other 18 would have favoured the punters. Every race had at least one runner whose victory would have cost Coral money.

Away from Dettori Day, punters have a keen interest in the general question: which horse offers better value to the punter, a 5:4 favourite or a 66:1 outsider? The 1978 Royal Commission considered this matter in some detail, and later studies by academic statisticians and economists have reached broadly similar views. Low-priced horses yield a better average rate of return than long shots.

There is no simple formula, but the Royal Commission found considerable stability over a period of 25 years: if every horse that started at 4:6 or shorter were backed, the punter could expect to see a small *profit* (before tax). If the SP odds were longer than 9:1 but shorter than 20:1, the average *loss* was about 35% of stakes. Backing horses priced at 20:1 or higher would lead to a loss of some 70% of stakes. Data for the 1973 flat season show that out of 4000 horses whose starting prices were either 33:1 or 50:1, just 23 were winners. The bookies could have offered 100:1 against all these horses, and still shown a comfortable profit. More recent data indicate that short-priced horses have been close to break-even, but once the odds offered are 10:1 or longer, average losses again tend to exceed 30%.

Several theories compete to explain why there is such a radically different rate of return on short-priced and long-priced horses. One idea draws a parallel with 'insider trading' on the Stock Exchange: some punters have knowledge, denied to others, that enables them to be almost certain which horse will win, and their money shortens the odds on winners. Another suggestion is to consider what punters and bookmakers expect of each other. Both parties willingly enter a risky game, knowing the dice are loaded towards the bookies. Bookies cannot make a profit unless they can tempt the punters to bet, so their prices must not be extortionate. Punters like to win, but they particularly relish the prospect of winning a sum large in relation to their stake. Even if objective belief is that the horse's winning chance is 1%, punters are attracted to the idea of winning as much as £33 for a £1 stake, and the bookies have negligible incentive to offer odds longer than 33:1. To do so would increase their liabilities, while attracting hardly any more money. Consider also the other extreme, where one horse is plainly superior to the rest, and might be expected to win at least two times in three. At odds of 1:2, a punter has to risk £50 to make a £25 profit, and few may be inclined to do this. In order to take any money at all, the bookie may need to make the bet more attractive, perhaps pushing the odds to 4:7 in the punters' favour. Betting on horses is not like throwing dice or tossing pennies: estimating the objective chances is an imprecise art. But bookies know that much money lost on short but over-generous prices in races with few runners will return to them when the fields, and the odds, are bigger.

The bookmaker is in more control when offering bets on soccer results. A typical example comes from all 66 matches on one coupon for a week-end in September 1997: the odds from match to match were quite variable, but every time produced a bookies' margin of between 11% and 13%. These coupons are distributed at the beginning of the week, which leaves several days for the uncertainties of injuries, transfer deals, etc. to have a significant effect on punters' perceptions. The main protection for the bookies is that bets must be made on a *combination* of matches. One rule is that if a punter wishes to include any home win, he must forecast at least five results, and all must be correct. The bookies' margin works like compound interest, so if the margin on a single match is 12%, the margin on a five-match combination rises to 76%! If the punter is prepared to bet on draws and aways only, he may combine just three matches, which still leads to a 40% margin. A better deal is the concession that when a match is being shown live on TV, bets on that match alone, or in combination with just one more match, are allowed.

(A cunning punter might be able to place a bet on what turns out to be a single match by taking advantage of the weather forecast. If blanket snow is expected in Lancashire, while London will be mild, the crafty combination of a top London game with several from lesser leagues, avoiding clubs with undersoil heating, could turn out well.)

After the National Lottery began, bookies hoped to be able to offer 'side betting' on the outcomes. Their intention was to accept bets, at fixed odds, either on single numbers or on combinations. Since the betting tax takes 9% of winnings, while Camelot pays 40% of its revenue to the government, the bookies expected to be able to offer attractive bets. The government came to the same conclusion, so took steps to prevent the use of National Lottery draws in this way, to protect its own tax take.

But there was no legal obstacle to the bookies using the results from overseas lotteries. Side betting in Ireland has long been permitted on the

Irish 6/42 Lottery, and the UK chains began to offer bets based on the twice-weekly Irish results. They then had the idea to set up their own daily fixed odds game, in the familiar 6/49 UK format. A team of bookies bought their own version of the Lottery machines and balls from the French manufacturers, and organized an independent firm of accountants to act as witnesses to verify the integrity of each draw.

Punters can select from a variety of bets, packaged to mimic combinations used at race meetings, such as 'Yankees' or 'Patents', or derived from the layout of Bingo cards. However, all these exotic wagers are based on the five basic bets of choosing either one, two, three, four, or five numbers from the list of 1 to 49. To win, *all* your numbers must appear in the winning set of six drawn that day. Table 11.2 shows the odds typically available, your winning chances, and what rate of return you can expect.

To illustrate how the calculations are made, consider the bet based on choosing three numbers, with payout odds of 511:1. Suppose your choice is 12, 31, and 41—the argument applies equally to any other set. The chance the six numbers drawn include your first number, 12, is plainly 6/49; when this happens, there are five other numbers chosen from the remaining 48, so the chance they also include your second number, 31, is 5/48; and similarly, the chance of also including your third number is 4/47. Taken together, the chance all three are included is

$$\frac{6}{49} \times \frac{5}{48} \times \frac{4}{47} = \frac{5}{4606},$$

less than 1 in 920. If you win, you receive a total of 512 times your stake, so the average return is

$$512 \times \frac{5}{4606} = 0.55579\ldots,$$

giving the figure of 55.6% in the table.

Table 11.2 The chances of winning, the odds paid, and the average return to the punter for the bets available on the "49s" game in the major chains' betting shops.

Number chosen	Winning probability, 1 in	Payout odds	Average return (%)
1	8.17	11:2	79.6
2	78.4	48:1	62.5
3	921.2	511:1	55.6
4	14 125	6560:1	46.4
5	317 814	99 999:1	31.8

The table shows how the average return decreases as the punter seeks more ambitious bets. In this game, the odds are quoted to include the tax, so a punter who successfully forecasts three winning numbers does receive £512 for a £1 stake. There is a cap of £250K applied for payment to any one customer or syndicate on any one draw, to protect the bookies' exposure to the risk of an enormous payout.

This helps explain why no bet is offered for forecasting all six numbers. Recall from Chapter 2 that the odds against achieving this are about one in 14 million. The last column in Table 11.2 suggests that, if such a bet were offered, the payout might drop to around 20%, indicating that odds of about 3 million to one should be offered. With the payout capped at a quarter of a million pounds, any winning bet of more than 8p would breach the limit! Bookies want neither the expense of handling losing bets of trivial amounts, nor the volatility in payouts from huge, but rare, wins.

Can you win from the bookies?

Some punters do so consistently. They need to have sufficient knowledge to be able to identify favourable bets, and enough self-control to bet at those times only. But even if you satisfy both criteria, there remains an important question: how much should you stake on any particular event?

Probabilists have known the answer since John L. Kelly Jr published it in 1956. Take the simplest case of betting in a favourable game: the bookie pays out even money on winning bets, but the chance of a win exceeds 50%. If you bet all your money, time after time, you will do well, on average. But this average is made up of a decreasing chance of having a rapidly expanding amount, and an increasing chance of the catastrophe of having zero. You have to be more cautious.

Kelly considered the strategy of betting a fixed fraction of your current fortune each time. In a favourable game, your fortune ought to grow exponentially, like compound interest. He worked out the way your rate of growth varies according to the fraction you bet. If you bet only a tiny fraction, you will not go bankrupt, but your fortune grows very slowly as you are investing too little; make the fraction too large, and the losses, when they occur, can be disastrous. Remember that a 50% wage rise, followed by a 50% wage reduction, leaves you 25% worse off than before! Kelly's answer is that the right balance is struck when the fraction you bet exactly measures the size of your advantage. If the chance of Heads is 51%, and Tails is 49%, the advantage is 2%: you should bet 2% of your fortune on Heads. Consider a bigger advantage—Heads is 55%, Tails is 45%: here, bet 10% of your fortune on Heads. This 'Kelly' strategy maximizes the long-term average growth rate.

You have to be patient. There will be fluctuations in your fortune, but you are relying on the long-term average. With the example of staking 2% of your capital when the chance of a win is 51%, this long-term average growth rate is only 0.02% each time. With the chance of a win up to 55%, the Kelly strategy gives a growth rate of about 0.5%. Seeking a bigger growth rate increases the chance of bankruptcy.

When the bet is advantageous, but the payout is not at even money, you need to scale the loss probability to find the Kelly strategy. Take the example where the bookie offers to pay at odds of 3:1 and your winning chance exceeds 25%. This is plainly favourable, but the right balance between a timid bet and a rash one is not at all obvious. To keep the arithmetic clean, suppose the chance you win is 28%. That means the chance you lose is 72%, and as the odds paid are 3:1, you must scale this by a factor of 3; and 72%/3 = 24%. Kelly's analysis shows that you should bet 28% − 24% = 4% of your capital each time.

Try this for other payouts. Suppose your winning chance is 34%, and the odds are 2:1, just favourable. Your losing chance is 66%, which, scaled by the payout factor of 2, becomes 33%; you should stake 34% − 33% = 1% of your capital each time to maximize your long-term average growth rate. The same method works with odds-on bets. Perhaps your true winning chance is 70%, and the payout odds are 1:2. Here your losing chance is 30%, which when divided by the odds of 1:2 becomes 60%. (Odds of 1:2 can be rewritten as odds of 1/2:1.) The Kelly strategy is to bet 70% − 60% = 10% of your fortune.

There are two important provisos. The first relates to the practicality of this advice. For example, suppose you have £1, and the recipe says you bet 10% of your fortune. If your first bet of 10p loses, you have 90p, and so will bet 9p next time. If that loses, you have 81p left, and so you seek to stake 8.1p. No such bet is possible. Even if your capital is far larger, you will often come across this problem. The Kelly strategy assumes you can chop up your capital into arbitrarily small amounts to meet the recipe. We need to modify it to cater for the impossibility of making bets such as 8.1p.

What you should do is to *round down*. Instead of 8.1p, you bet 8p. If that wins at even money, you now have 89p; you would like to bet 8.9p, which is not permitted, so you round down to 8p instead. There might come a time when you have less than 10p, which calls for a bet of under 1p, which you round down to zero. Tough. Even in favourable games, an unlucky streak can happen. The casinos in Monte Carlo operate favourable games, but once in a blue-ish moon, someone breaks their bank.

The other proviso stems from this chance of a long losing run. The Kelly strategy achieves the best possible long-term average growth rate, but some people are, quite understandably, cautious. They are prepared to sacrifice part of this potential growth for a greater assurance of avoiding a disaster. The way to do this is consistently to bet a little *less* than the Kelly fraction. As you are in a favourable game, your fortune will rise in the long run, but more slowly. In recompense, you are less likely to face ruin. It was this consideration that led to the recommendation to round down both 8.1p and 8.9p to 8p. Betting slightly more than the Kelly fraction will both reduce your long-run growth rate, and increase your chance of hitting Carey Street. Betting a little bit less than the Kelly fraction is superior to betting that same little bit more.

In a race with many horses, there may be more than one bet that you see as favourable. Perhaps there are two horses, priced at 4:1 and 5:1, both of which seem to you to have a 25% chance of winning. Betting on either places you in a favourable game. Kelly's work also covered this situation; he showed that your best move is to bet appropriate amounts on both horses. He gave a general method of calculating how much to stake, and what that would mean for the overall average growth rate of your capital. One of his conclusions will surprise many gamblers: sometimes, you ought to include in your bets horses whose prices are *unfavourable* to you.

That statement runs counter to intuition, but it flows logically from Kelly's analysis. Take the example of a race with five horses, A, B, C, D, and E, with odds of 13:8, 9:4, 9:2, 5:1, and 20:1 on offer. (This has a bookies' margin of about 8%.) Suppose, after poring over the form books, you assess the five winning chances as 40%, 30%, 20%, 8%, and 2%. You have them in the same order as the bookie, but there are two horses where the odds are in your favour. You think C will win one time in five, and the price of 9:2 is better than that; and similarly the odds of 13:8 for A are better than you expect if it wins 40% of the time. You see betting on B, D, or E as losing propositions, as the odds on offer do not match up to their winning chances. Nevertheless, if you have the standard aim of making your fortune grow as quickly as possible, your best strategy is to bet on a suitable combination of C, A, *and also B.*

Kelly tells you how to find this best combination. It needs more mathematical symbols than I wish to display here, so you might consult Appendix V. Here I will simply report the results of the Kelly arithmetic for this scenario. You should bet nearly 23% of your capital: split it 10.6% on A, 6% on C, and 6.25% on B. That will lead to an average growth rate of 0.35%, so if you could repeat this a large number, N, times, you could be looking at a fortune about $(1.0035)^N$ times as big as what you started with.

But suppose horse B had been on offer at the worse odds of 2:1, with A and C at the same prices, and D, E lengthened slightly in compensation. This time, the Kelly analysis omits horse B, and insists you bet on A and C only. Your best move is 5.15% of your fortune on A, and 3.37% on C, which leads to a long-term average growth rate of 0.21%. It should be no surprise that you are worse off: one of the better horses has had its price cut, but the reduction in the growth rate is substantial.

This example demonstrates why the counter-intuitive move of including horse B in your betting strategy in the first scenario begins to make sense. Including B leads to you investing nearly 23% of your capital, at favourable odds *overall*. Horses A, B, and C will win 90% of all races between them, so you lose all your bets only one race in ten. But when the odds dropped to 2:1, and the Kelly strategy excluded B, recall that A or C will win the race only 60% of the time. You have to be more cautious in your betting policy, and commit only 8.5% of your capital. The overall odds are better, but 91.5% of your capital is sitting idly by, in reserve against disaster.

Betting on B alone, at any odds worse than 7:3, is a losing strategy. But the odds of 9:4 turn out to be close enough to 7:3 for it to be sensible to bet on B, so long as you also have the right amounts on A and C.

The Tote

The Tote, as the Horserace Totalizator Board is known, is a statutory authority established in 1929 with the twin purposes of providing a different form of betting, and raising money to support racing. Betting on the Tote is parallel with entering the Football Pools, in that a proportion of all money staked is deducted to cover expenses and profits, with the rest returned to the winners in proportion to the amounts they bet. Thus a punter who bets on the Tote is competing against the other punters. Greyhound racing also operates Tote pools on the same principles, and in some other countries, the pari-mutuel is equivalent to the UK Tote. Any

Tote operation is a no-risk business, as it never pays out to punters more than has been staked.

Suppose Bob wishes to place a Tote bet on the next race, which has ten runners. He discards the idea of the dual forecast, which asks him to name the first two, in either order, and looks for a likely winner. Above the betting window is a television screen that is continuously updated as the bets from other punters are recorded. This indicates what the payouts would be, if the race started now. In the early stages of betting, these payouts will vary rapidly, so Bob waits until just a few minutes before the race to see which horse gives best value. A cautious man, he bets £2 'each way' on Beetle Bomb.

This is two separate bets: £2 on the horse to win, and £2 on it to be placed in the first three. When betting closes, perhaps a total of £5000 has been bet on the different horses winning, and £3900 on the places. The Tote deducts 16% from the win total, and 24% from the place total. This leaves £4200 to pay out in win dividends, and £2964 for the places, in separate pools. Beetle Bomb wins.

The amount Bob receives depends entirely on how much had been staked on that horse. Suppose this sum was £1000. There is £4200 to share out, so each £1 bet qualifies for £4.20p; Bob bet £2, so collects £8.40p. If only £100 had been bet on Beetle Bomb to win, the win dividend would have been £42, and Bob's share £84.

For the places, the first step is to divide the £2964 place pool equally across the horses that finished first, second and third; no distinction is made, so £988 belongs to the Beetle Bomb place pool. The same steps are followed. If £500 had been bet on Bob's horse, the return to a £1 stake would be £(988/500), which would be rounded down to £1.90p. Bob collects another £3.80p for his £2 place bet. But if the total staked on the horse had been only £100, the return to £1 would be £9.80p. The payouts on the horses placed second and third could be higher or lower than for Beetle Bomb.

There is scope for anomalies. With the totals as given here, suppose £1000 had been bet on Bob's horse to win, but only £100 on it to be placed. The rate of return to the win bets is £4.20p, but the return to place bets on the same horse rises to £9.80p! Famously, something like this did happen in the 1973 Belmont Stakes in the USA. The 'wonder horse' Secretariat was priced at 10:1 *on* by the bookmakers, and duly trounced the field, winning by 31 lengths. But on the pari-mutuel, the return to a $2 stake was $2.20 for a win, and yet $2.40 for a place! Punters presumably saw Secretariat's victory as a foregone conclusion and shunned the unappealing win odds. Instead, they spread their 'place' money over the field, overlooking the fact that one-third of the place money would be allocated to those backing the winner. Place betting is possible when there are five or more runners: with five, six, or seven horses, just the first two are considered placed; with eight to 15 runners the first three; and with 16 or more it is normally the first four.

On these Win or Place bets, the rates of return to the punter are 84% and 76% respectively. On the other Tote bets, the rate ranges from 71% to 74%. One bet is the jackpot, the challenge to name all the winners of the first six races; if no punter succeeds, the jackpot is carried over to another day. Just as with a rollover in the National Lottery, this injection of free money from elsewhere stimulates more punters to chance their luck.

A Tote win dividend of £12 is plainly equivalent to odds of 11:1. Punters who make bets in off-course betting shops may ask to be paid their winnings at 'Tote odds', thereby automatically controlling the size of the bookies' margin. The Tote organization earns extra income by charging other bookies for the right to use these odds. Since the Tote also owns a chain of bookmakers' shops, the bookies argued that, in equity, the Tote ought to pay for the right to use the SPs. But as the Royal Commission pointed out, the bookies merely supply the raw material for the *sporting journalists* to derive appropriate SPs! The Tote gets its SPs free.

Spread betting

Spread betting, or index betting, is a recent and rapidly expanding form of organized gambling. It operates in the way the stock market trades the future prices of commodities. For delivery of a consignment of coffee beans in six months time, a spread of £1500–£1600 may be quoted. The lower price is what a producer would receive, the higher price is what a consumer must pay. Contracts to purchase or supply coffee beans at these prices are themselves tradable articles, and profits or losses are made according to the difference between the contract price agreed now, and the open market price at the times the beans are delivered. Instead of coffee beans at monetary prices, spread betting on sports events is based on how well teams or individuals perform against upper and lower limits set by the index betting firm.

An illustration of this concept is the estimate of the number of runs the

Australian cricket team may make in the first innings of a match against England. The spread may be set at 340–360, and punters have a choice of two types of bet. Megan believes that Australia will do badly, so decides to *sell* at £2 per run. The selling price is 340, and every run short of 340 gains the bettor £2; so if Australia make 300 runs, Megan gains £80. But if Australia made 450 runs, that is a disaster for her: that score is 110 more than the selling price, which means she loses £220. Martin fancies Australia to do well, and opts to *buy* at £3 per run. As the buying price is 360, an Australian score of 450 would yield him a £270 profit, while a score of 300 leads to a loss of £180. Table 11.3 illustrates their conflicting fortunes, according to Australia's score.

Like bookmakers offering fixed odds, the index betting firms are backing their judgement against that of the punter. There are three ways in which the index firm can look to make a profit. First, if the actual result falls within the interval covered by the spread, the firm wins every bet. In this cricket example, with the spread of 340–360, suppose Australia scored 345; then every seller has lost five times the unit stake, and every buyer has lost 15 times the unit stake.

The second way is to make the bet equally attractive to buyers and sellers. If exactly the same amount of money—£1000, say—is staked on both buy and sell, the firm automatically makes a profit of £1000 × (the size of the spread). Just as bookmakers will usually adjust their odds in response to the sums staked on different horses, so an index firm may change its original view if the punters bet heavily one way.

The third way is by being better than the punters at forecasting the outcomes. In some matches, a spread of 340–360 runs may look a very inviting buy, and three times as many punters will buy as sell. One possible response is to edge the spread up to 390–410 to deter buyers, hoping to equalize buy and sell stakes. The 'courageous' response is to leave matters alone, and welcome more buyers, gambling that these punters are wrong. If that judgement were correct, and the team are dismissed for a low score, profits would be substantial. But the judgement could be wrong. Barings'

Table 11.3 The spread for the Australian score is 340-360. Megan "sells" at £2 per run, Martin "buys" at £3 per run.

Australian score	250	300	350	400	450
Megan's profit (£)	180	80	−20	−120	−220
Martin's profit (£)	−330	−180	−30	120	270

Bank collapsed through slack supervision of one trader, whose intuition about the way the markets would behave was faulty.

Just as bookmakers face the problem of how to set initial odds to attract bets across the field, so index firms have to find a spread that will interest both sell and buy punters. The firm knows it will not win money on every event, but it intends to be in business for a considerable time, so it concentrates on *averages*. In the long run, averages rule. What advice should be given to a firm on a suitable spread for the time of the first goal in a top soccer match?

Currently, a match average of rather more than two goals, but a little less than three, is about right. Given the number of goals, *when* do they occur? Statistics show that, in the main, goals come along fairly randomly, with a tendency for the rate to increase a little as the match goes on. If goals are scored at random, at a given average rate, probability considerations lead to a neat answer: the average time to the first goal is just the reciprocal of the scoring rate. Suppose 2.6 goals are scored, on average, over 90 minutes. Then the average time to the first goal is not far from $90/2.6 = 34.6$ minutes. This might be adjusted up to about 36 minutes, as later goals come a little quicker. Finally, modify this average time to take account of the particular teams: higher if fewer than 2.6 goals are expected, lower if both teams are attack-minded.

Having estimated the average, the spread should centre around this value. *An index betting firm that sets its spreads to enclose the average will take money from punters in the long run.* This is because, on average, the firm gains from both buyers and sellers. The width of the spread is a separate question, related to the risk the firm is willing to take. A cautious firm that sets a wide spread will make few losses, but do hardly any business. We noted that if sell and buy totals are equal, the firm's profit will be this total multiplied by the width of the spread. So to make it worth while to halve the spread, at least twice as many punters must be attracted.

In many events, the spread is continuously updated while the game takes place, and further bets can be placed. This gives punters the chance to lock in to guarantee a profit, or to cap their loss. Continue the example of the Australian cricket total with an initial spread of 340–360. On a flat Oval pitch with the tea-time score at 220 for two wickets, the spread may have moved to 420–440. Luke, who had bought Australia initially at £2 a run above 360 could now sell at £2 a run below 420. This makes certain that he has a profit of £120, whatever the Australian score! Try some examples to convince yourself: if Australia collapse, and make only 260, he *loses* £200 on his first bet, but *wins* £320 from his second; should Australia press on to make 500, his first bet wins £280, his second loses £160. The net outcome is always the same. On a different day, Australia might be 160 for five wickets, with the spread now quoted at 230–245; Luke, who bought Australia at £2 per run, now fears a collapse. He decides to cut his loss by selling at £2 a run. Whatever the total, he loses £260, but if Australia scored only 180, he has saved himself a loss of a further £100. The activity of betting twice, in opposite directions, is known as 'closing' a bet. It can be settled now, as future events have no effect on the outcome.

One obvious danger in index betting is that a punter may not know the maximum loss he can suffer, as he does not commit a definite stake. In the cricket example, a punter who sells at £2 a run may watch in agony as Australia bat on into the third day, taking the score above 700, each leg bye costing another £2. But also the gain for a punter who had bought Australia is unlimited. Just as with stock market investments, advertising by spread betting firms contains a 'wealth warning' that markets can be volatile, and move rapidly against you. Sometimes, an upper limit can be set on losses and gains, but punters should be aware of the extent of their risk.

Spread betting on a single soccer match ranges from the obvious to the highly speculative. The most popular bets are on the total number of goals scored, and on the supremacy of one team over the other. Both of these are commonly expressed in tenths of a goal: Liverpool's supremacy over Aston Villa may be quoted as $0.6 - 0.9$ at the start of a match, and if they scored two early goals, it might move to $2.7 - 3.0$. Each player has a known shirt number: you can bet on the sum of the shirt numbers of all the team's goalscorers! Another artificial bet is on a performance index for a team: collect 25 points for each goal scored, 3 for each corner won, and 1 for each minute a clean sheet is kept; subtract 10 for each goal conceded, 10 for each yellow card received, and 25 for each red card. And good luck as you consider the spread of 45–50, and wonder which way to go!

Some Performance Index is in common use in knockout competitions: the winner may be awarded 100 points, the losing finalist 75, the losing semi-finalists 50, and so on. When there are just two teams left, the spread has a complete analogue with the margin used by bookies. In the match for the final of the Rugby League Premiership in 1997, the quote for Wigan was 92–94, that for St Helens was 81–83. To buy Wigan at £1 is to win £6 if they are successful, but to lose £19 if they lose: bookies' odds would be 6/19. Similarly, the odds for St Helens would be 17/8, giving the sum of the odds as $19/25 + 8/25$, a margin of 2/25, or 8%. The formula for the margin is simple for this two-horse race: it is the ratio of the width of the spread to the difference in the index between winning and losing $((94-92)/(100-75))$.

Bets available on an afternoon's horse racing include: the supremacy, in lengths, of one named horse over another; a performance index on all the favourites at a meeting—score 25 points each time a favourite wins, 10 points if it is second, 5 if it is third; Stop at a Winner starts with ten points, and adds ten points each race until the first time the favourite wins. Other bets can be based on adding up the SPs of all the winning horses; a performance index for a named jockey; the sum of the ages of all the owners of the winning horses (I made that last one up, but it seems no more outlandish than several bets that *are* on offer).

A common system gives 50, 30, 20 and 10 points respectively to the first four horses, and some spread is offered for each horse. This means that a punter can back any horse to do well, or to do badly—an option not usually available with odds bets. Since there are $50 + 30 + 20 + 10 = 110$ points that will be scored, the smaller numbers in each spread must sum to at most 110, and the larger ones to at least 110. For otherwise, by selling every horse in the first case, or buying all of them in the second case, a punter could guarantee a profit, whatever the race outcome. To guard against this blunder, the sum of the centre points of the spreads will be close to 110.

How should a firm set about constructing spreads for a given race? One way is first to estimate each horse's chances of being first, second, third or fourth. This leads to an estimate of the average score for each horse, which can be taken as the midpoint of its spread. The widths of the spreads will govern the firm's margin. The wider the spreads, the less risk to the firm, but very wide spreads could deter punters from betting at all. After a set of initial spreads has been published, the spread firms will adjust them to reflect punters' interest, just as Coral changed the odds on Dettori Day, but they have to begin somewhere.

There is no simple analogy in this set of spread bets with the overround for a set of definite odds that we calculated earlier, although suggestions have been made. My own approach (for details and reasoning, see the reference at the end of this chapter) is to compute the ratio of two quantities: the numerator is the sum of the widths of all the spreads, the denominator is twice the difference between the scores of the winner and the horse that finishes last. So, for example, in the seven horse race of Table 11.1, perhaps the spread for the favourite would be (27, 29), four other horses might also have a spread that is two points wide, and the two outsiders could be offered at (6, 7.5) and (3, 4.5). The numerator is then

$5 \times 2 + 2 \times 1.5 = 13$, while the denominator is $2 \times (50-0) = 100$, leading to a spread betting "margin" of $13/100 = 13\%$.

Firms that offer both fixed odds and spread bets on the same race must ensure their prices are reasonably equivalent, or again a shrewd punter may be able to exploit the differences to guarantee a profit.

American football is a splendid sport for index betting; two popular bets are based on the points difference between the two teams, and the total points scored in a game. The data indicate that a little over 40 points are scored per game, on average; typical spreads for the total points in a single match range from 37–40 to 44–47 across a dozen games. One weekend in 1997, the quote for the total number of points over all 12 matches was 500–510. At first sight, in comparison with the spread on a single match, this looks surprisingly narrow. It is ten points wide, the usual spread on an individual game is three points. But there is a straightforward statistical reason that reconciles these two figures.

It is no surprise to see the middle of the spread for 12 matches at 505 points, which corresponds to about 42 points per match, quite consistent with the average for the season. The width of any spread will be associated with the variability of the quantity of interest: the greater the variability, the wider the spread. One statistical way to measure variability is the standard deviation (Appendix III). Recall that when independent quantities, such as the points scored in different matches, are added together, the *variance* of this total will be the sum of the individual variances, and that the standard deviation is the square root of the variance. So the standard deviation of the total score in all 12 games will not be 12 times as big, it will be the square root of this. And $\sqrt{12} = 3.46..$, which suggests that the spread of the total over these matches should be between three and four times the spread for a single game. So it is—compare 500–510 with 40–43.

That same weekend saw 497 points scored over the 12 games, so the sellers won just three times their unit stake, while the buyers lost 13 times theirs. If one converted touchdown had been disallowed, reducing the total points scored to 490, the sellers would have won ten times their stake, the buyers would have lost 20.

To tempt gamblers, the spread betting firms often offer a combination bet over a number of events taking place on one day, or over a weekend. The following example relates to late 1997. Many of the events were being shown live on satellite TV, and punters could telephone bets 'in the running' as the spreads were updated. I. G. Index invited bets on a single number, the total of the following quantities:

(1) The number of points by Leicester in a Rugby Union match.

(2) One point for each completed minute in a Lennox Lewis boxing match.

(3) 20 points for each goal scored by Middlesborough in a soccer match.

(4) Similarly, 20 points for each Chelsea goal in a different match.

(5) The number of points scored by Kansas in an American football game.

(6) Ten points for each shot under par in the final round by the golfer Davis Love III.

The quoted spread was 155–165; the six individual results were 22, 2, 20, 40, 14, and 40, giving a total of 138. After the first three events had given a total of 44, the spread on offer was 124–130—narrower than the original to reflect the reduced uncertainty with just three events left.

Waiting for the spread firms to blunder, and offer you a guaranteed winning bet is as unproductive as waiting for the bookies to construct a negative margin. But the firms are so catholic in their tastes, and anxious to attract bets, that a knowledgeable sports fan can hope to catch them out. In the 1996 European soccer championship, it was known that 31 matches would be played. One spread on offer was for the fastest goal in any of these matches; the initial quote was 230–245 seconds.

I claim that bet is a clear sell. For *none* of 31 games to produce a goal within 230 seconds is equivalent to 118 minutes without a goal! The odds are heavily against that. Indeed, several matches did produce a very early goal, the fastest being inside two minutes. Risk-averse that I am, my money stayed in my pocket.

Postscript (November 1999)

(i) the fear expressed on page 216, that a high betting tax could lead to attempts to avoid it, is being realised. With the growth of telephone betting, and the use of the internet to advertise prices, some firms have set up overseas operations with a "levy" on bets of just 3%, well below the 9% that operates in the UK mainland. Two government departments—the Treasury, who wish to maximise tax income, and the Home Office, concerned to regulate gambling—may have a conflict of interest. But the punters and the bookmakers are on the same side!

(ii) My explanation of how a "margin" might be calculated for a horse race in which definite "rewards" are given according to where the horses finish can be found in *The Statistician* (1999) "(Performance) Index betting and fixed odds", Vol 48(3) pages 425–34.

....

CHAPTER 12

This sporting life

Thank goodness the old boxing saw 'A good big 'un will always beat a good little 'un' is false. Sport would attract far less attention if it were more predictable. Tennis matches turn on the effect of a net cord, or the exact distance a lob from 20 yards will travel. The capricious nature of decisions by officials—whether the ball really crossed the line, why that try was disallowed—affect outcomes in profound yet random ways. This chapter explores a number of places where an investigation of the effects of chance might lead to a better understanding of the game, or possibly point to better tactics.

Soccer

A feature of soccer is that goals are rare. Even the best teams at top level average under two goals during a match of 90 minutes. This fact is enough to make soccer results far less predictable than those in many other sports, and reminds us why clubs recognize the season-long League championship as more prestigious than a knockout Cup. One unlucky day, and the Cup run is over.

The simplest plausible model of the scoring assumes that goals occur at essentially random times, at some average rate that depends on the team, its opponents, and who is playing at home. If this is a reasonable model, then the actual number of goals scored by a given team in a match would follow a Poisson distribution (Appendix II) with the appropriate average. When statisticians investigate soccer matches, they find this model is pretty good, although there are ways to improve it: for example, after the first goal has been scored, later goals tend to come along a little quicker. But a model that assumes a Poisson distribution is adequate for many purposes. To help get a feel for what this means in practice, Table 12.1 shows the (Poisson) frequency with which a given team will score 0, 1, 2, ... goals in a match. Even teams who average two goals a game fail to score

Table 12.1 Assuming the number of goals has a Poisson distribution with a given average, the probabilities of a team scoring the number of goals indicated.

Goals	0	1	2	3	4 or more
Average = 0.8	45%	36%	14%	4%	1%
Average = 1.2	30%	36%	22%	9%	3%
Average = 1.6	20%	32%	26%	14%	8%
Average = 2.0	14%	27%	27%	18%	14%

about one match in seven. Whatever the average, there is some most likely number of goals, with the probabilities of more goals or fewer decreasing as we move away from that central number.

We can use such frequencies to estimate the chance that either side will win. Take a game in which the home side is expected to score 1.6 goals, on average, while the corresponding number for the away side is 1.2. This is fairly typical of closely matched teams at the top level. Assuming the numbers of goals scored by each team are independent, the probability that the game ends in any particular score is then easily found. For example, using Table 12.1, the chance these teams play a 0–0 draw is (20%) × (30%) = 6%. The chance of a 1–1 draw would similarly be (32%) × (36%) = 11.5%. By considering each score that leads to a draw, and adding up the corresponding chances, we estimate that the chance this match ends in a draw is about 25%.

A similar calculation, this time aggregating the probabilities of all scores that lead to a home win, indicates that about 48% of such matches end this way, which leaves 27% as the chance of an away win. Estimates calculated in this fashion match the real data quite well. A team that would be expected to 'lose' by 1.6 goals to 1.2 goals will win over a quarter of such encounters. Bookmaking firms use arguments similar to the one described here to assess the chances of the three possible outcomes, and so set odds to tempt punters. If you are better than the firms at estimating the average number of goals a team would be expected to score, you could use these methods to calculate the corresponding winning chances, and perhaps identify some good bets.

The first goal

How important is it that a team scores the first goal? If that goal occurs early in the game, the opposition has plenty of time to strike back, but a first goal in the 88th minute is almost always enough for victory. Leaving aside *when* that first goal is scored, what is a team's overall chance of winning, given that it scores first?

If the teams are ill matched, it is overwhelmingly likely that the better team will both score first and go on to win. There are exceptions, of course. In 1993, San Marino took the lead in the first few seconds, but England won 7–1 in the end; and sometimes—USA 1, England 0, World Cup 1950—the underdogs score first and hang on. But in these mismatches the chance that the team who scores first wins the game is over 95%. It is more interesting to consider games between equally matched teams.

In such a game, if there is only one goal, it is plain that the team that scores first must win. So look at games in which exactly two goals are scored. After the first goal, either team is equally likely to score the other one, so half the time there will be a 1–1 draw, and half the time the team that scored first wins 2–0. In games with exactly three goals, a draw is not possible. The only way in which the team that scores first fails to win is if both the other two goals are scored by the opposition. Since either team has probability one-half of scoring any particular goal, the probability the opposition score both the last two goals is one-quarter. In these matches, the team that scores first loses 25% of the time, and wins the remaining 75%.

One final calculation before we bring it all together: look at games in which a total of four goals are scored. Label goals from the team that scored the first goal as F, and from the other team as O. After the initial goal, the remaining three goals can be split in eight equally likely ways

FFF FFO FOF OFF FOO OFO OOF OOO.

The first four lead to victory by the team who scored first, the next three lead to a draw, and the other team wins in the last case. The respective chances are one-half, three-eighths, and one-eighth.

Plainly, for any fixed number of goals, a similar analysis can be made. If there is an odd number of goals, a draw never occurs, and the chance that the team who score first go on to win always exceeds 50%. It is not difficult, using the methods of Appendix II, to write down this probability. When there is an even number of goals, at least four, all three results are possible. Exactly half the time the team who scores first will win; again, it is not difficult to write down the chances of a draw and a loss.

To give an overall answer, we need to know the frequencies of one, two, three, etc. goals within a match. Goal-less draws are eliminated from our calculations. We have noted that the Poisson distribution generally gives a good model for the number of goals scored about a particular average, so now we simply scale up the probabilities in a Poisson distribution to take account of the removal of 0–0 scores.

Over a wide range of professional competitions, the average number of goals per game lies in the range 2 to 3.5; a value of 2.8 is reasonably typical. There is one calculation of pleasing consistency: when 0–0 draws are excluded, the chance of a drawn match is very close to 20% over that whole range from an average of 2 goals per game up to 3.5. Against a team of equal strength, you draw about 20% of matches in which you score first.

The fewer goals are scored, on average, the more important it is to get the first one. Weighting the different numbers of goals according to the frequency with which that number is scored, you win 72% of matches in which you score first if the average number of goals is as low as 2. Since 20% of these games are drawn, you lose here only 8% of the time. When the average number of goals is as high as 3.5 per game, your chance of converting an opening score to a win is about 64%, so you lose 16% of the time. Taking the middle value of 2.8 goals per game, the team that scores first wins 67% and loses 13%, or so.

Recall that these calculations assume the teams are well matched. Where one team is measurably superior, we expect the team that scores first to go on to win a little more often than these figures indicate. Within a single League, the top teams will be noticeably better than those near the bottom, so for a season's data, and some 2.8 goals per game, we might expect the team that scores first to win some 70% of the time, to lose about 10% of games, with 20% drawn. In November 1997, the *Independent* newspaper gave accumulated figures over all the matches in the English Premier League over two and a bit seasons: of 825 matches in which at least one goal was scored, the team that scored first won 68.7%, drew 20.8%, and lost 10.4% of games. Statistically, the fit of these data to our model (as judged by Appendix IV) is excellent. This helps justify the use of a Poisson model for the number of goals in other circumstances.

Scoring the first goal always helps, but if you fall behind, take comfort from the game in December 1957 when ten-man Charlton Athletic trailed Huddersfield Town 5–1. At the final whistle, Charlton had won 7–6, leaving Huddersfield with an unusual record: six goals, all by different players, and still losing.

Yellow cards

Often, yellow cards (YCs) are issued by referees as a means of reminding players to stick to the laws. Generally, a YC is successful in that the player who receives one remembers not to infringe seriously again, or he will collect a second YC, and so be sent off. When a player has collected several YCs from different matches, he faces automatic suspension from a future game. This can be quite traumatic for the individual (recall Gazza's tears in the 1990 World Cup?), but it may be even worse for the team, by reducing team selection options. Sometimes, players may have to be chosen out of position if too many of the squad are suspended. Not many teams have the luxury of a Duncan Edwards or John Charles who could play almost any position with distinction. If a player has probability p of collecting a single YC in any match, for what proportion of games will he be suspended?

In World Cup and other major competitions, a common rule is that after a second YC, the player is suspended for one match, but then his YC count reverts to zero. That means that we can sum up his status at the end of any match by the label 0, 1, or 2, according as to how many current YCs he holds. Any player with label 2 misses the next match, after which his label is 0. The natural way for a probabilist to analyse this problem is through what is known as a Markov chain, which is described in the next chapter. But without using any new terminology, proceed as follows.

The chance of moving from label 0 to label 1 in a match is plainly the same as the chance of moving from label 1 to label 2, so in the long run a player has labels 0 and 1 equally often. His label changes from 1 to 2 in a match with probability p, so the chance his label is 2 is just p times the chance his label is 1. Since the three chances add up to 1, the only values consistent with these statements are that labels 0 and 1 have probability $1/(2+p)$, while 2 has probability $p/(2+p)$. Thus a player whose chance of picking up a YC in any match is p will be suspended from a fraction $p/(2+p)$ of his team's games. A careful player, with $p = 10\%$, will miss just one match in 21; a mid-field hatchet man, who tends to collect a YC every other game, misses one match in five. Coaches should remember these calculations when they name their squads.

Red cards

It is now accepted that if a defender commits an offence that denies an attacker a clear goal-scoring opportunity, the offending player receives an immediate red card and is sent off. Leave aside the effect of suspension in future matches, and look at the consequences within the current match. A team down to ten men is at a permanent disadvantage, but it may be better to save a certain goal a few minutes from the end, even at the expense of the sending off, if the team would lose otherwise. Professional teams do not follow the example of the famous Corinthians, who, if a penalty kick was awarded against them, instructed their goalkeeper to stand aside to ensure a goal was scored. A coach will expect his players to commit a 'professional foul' if that appears to be in the team's best interests. An analysis by G. Ridder and two colleagues now makes it possible to quantify this advice.

They looked at data from 340 matches in the Netherlands, where just one team had a man sent off, to assess what difference this made. They noted the *time* of the red card, the current score, and the final score. It is no surprise to find that the longer a team had to play with ten men, the worse they did, on average. There was no inevitability about the outcome—some teams do play better with ten men—but the overall message was clear. Ridder and his colleagues calculated the average extra number of goals that would result from the sending-off, and used the Poisson model to work out the chances of 0, 1, 2, etc. extra goals actually being scored. This led to the following cynical advice, in a game between two evenly matched teams. (Your aim is simply to find that action that minimizes the chance you lose the game.)

A defender, faced with the choice between an illegal act that will get him sent off, and allowing a clear goal-scoring chance, should make two calculations. First, he must assess the chance the opponent will score if allowed the opportunity; second, he must know how much time is left in the match. Balancing these calculations tells him what to do. The earlier in the game the dilemma arises, the more certain a defender must be that a goal will result, to make committing the offence worth while. For any particular probability of a goal, there is some crossover time: earlier than this time, he should *not* commit the professional foul, later than this time the team can expect to be better off if he plays unfairly and is sent off. According to the Dutch calculations, sample answers are:

Chance of a goal	Crossover time
100%	16 minutes
60%	48 minutes
30%	71 minutes

All these calculations are for offences *outside* the penalty area. If the offence is committed inside the area, in addition to the sending off a penalty kick will be awarded, and so the opponents have a very high chance of scoring in any case. Thus, about the only circumstances where it is better for a team to concede both a penalty kick and a red card are when a goal that would lead to defeat is certain, and the match is almost over. If the scores are level in the last minute, you *should* handle the ball on the goal line; but probably not with 15 minutes left.

This analysis applies when the two teams are well matched. The weaker team has a greater incentive to commit red card offences: if the opposition are better than you, the crossover times are earlier in the match. Conversely, if you are winning comfortably, there is greater risk from the cynical foul than from allowing the attacker his chance of a consolation goal.

Golf

Most professional tournaments are conducted as stroke play over 72 holes. The top players are so closely matched that one disastrous hole—taking eight on a par four—can effectively end a player's winning chances. By contrast, in match play, one disastrous hole is less serious; if the opponent scores four, it is irrelevant whether you take five shots or 20. The Ryder Cup between Europe and the USA is conducted at match play, and one particular component is the fourball. Each side has two players, and at each hole the better score of each side determines the outcome. In this form of golf, *inconsistency* can be a considerable advantage.

To see how much of an advantage, we make some reasonable assumptions about what good players are capable of, and explore where the model leads. At any hole, we take it that a player scores either eagle, birdie, par, or bogey, and we have set the data up so that the average score is about two under par. For two contrasts in consistency, I introduce Joe Par and Dave Erratic, whose respective behaviours at any hole are described by

	Eagle	Birdie	Par	Bogey
Joe Par	5%	10%	80%	5%
Dave Erratic	10%	20%	40%	30%

You can calculate that, on average, steady Joe Par does marginally better.

In a singles match between these two players, we can assess what will happen by examining each combination of scores in turn, and noting the frequencies. When the arithmetic is over, it turns out that they take the same score at any hole 36% of the time; Joe wins about 35.5% of holes,

Dave wins about 28.5% of holes. Joe has a small but definite advantage, and the best guess is that Joe wins an 18-hole match by about one hole.

But consider the three possible fourball pairings: two who play like Joe, two who play like Dave, and one of each type. Make the reasonable (?) assumption that all players' scores at any hole are independent. The distribution of the better ball score at any hole for each pairing can be calculated as

	Eagle	Birdie	Par	Bogey
Two Joes	9.75%	18%	72%	0.25%
Two Daves	19%	32%	40%	9%
Joe + Dave	14.5%	26%	58%	1.5%

The main effect has been to reduce the frequency of a bogey score from 30% to 9% when two Daves play together, and also to enable that pair to do better than par more than half the time. As a result, the pairing of these two inconsistent players makes them clear favourites to win a match against either of the other two pairings. For consider setting two Daves against two Joes; calculations show that the erratic pair will win 41% of holes, 36% will be shared, and the Joes win only 23% of holes. The Daves ought to win an 18-hole match by about 2 and 1, or 3 and 2. A match between two Daves and a Dave + a Joe is closer: the Daves win 36% of holes, the others win 30%. Finally, a Joe and a Dave would win 32% of holes against two Joes, who would win only one hole in five.

Another place where the inconsistent player may well triumph is in a 'skins' game. Here, a group of players contribute a sum of money to make a prize for each hole, and play a round together. If no player wins the hole outright by taking fewer shots than *all* the others, the prize is carried forward to the next hole. However, if one player does win the hole outright, he or she takes ('skins') all the accumulated money. In a skins match with four steady players, a fifth who oscillates between brilliance and mediocrity will rely on the others to cancel each other out when he double bogeys into trees and lakes, and expect to snatch several large prizes with his outrageous eagles.

As an aside from golf, note the possible value of inconsistency in certain events at athletics. In the long jump and triple jump, or the throwing events (javelin, hammer, shot, discus), contestants are allowed a number of attempts, with only the best one counting. Smith may do better than Jones on average, but one effort from Jones may take the gold medal. To put flesh on this bald statement, suppose Jan and Steve throw the hammer; any throw from Jan travels a distance having a Normal distribution (Appendix II) with mean 75 metres, and variance 16; Steve's average distance is 77 metres, variance 4. Steve throws further, on average, and is more consistent. But the more throws they are allowed, the more likely it is that Jan will produce the longest throw.

If just one throw were permitted, the chance that Steve wins can be calculated from tables to be about 67%. But in a standard contest where each is allowed six throws, Jan is favourite. Comparing his best with Steve's best, Jan's winning chance is about 58%. If each were allowed 20 throws, Jan would win about 75% of the time. Jan is far more likely than Steve to hold the world record, as that is decided on each person's lifetime best performance.

Cricket

For many years, the place of statistics in cricket was confined to recording the scores, evaluating averages, and being the source of trivial information ('When was the last time two left-handed brothers scored 50s before lunch at Lords?'). But recently, cricket authorities have accepted that statisticians have more to offer. Frank Duckworth and Tony Lewis devised an ingenious solution to a problem that had bothered the game for 30 years: find an equitable way to deal with rain or other interruptions that prevent a limited-overs match being completed according to the original schedule.

To present the problem at its simplest, suppose the match is scheduled for 50 overs. One team completes its innings as normal, scoring 200 runs, but then rain arrives and leaves time for just 25 overs for the team batting second. How should the match be completed?

Various solutions have been tried. At first, targets were simply adjusted pro rata. In the above example, as the second team could face only half the expected number of overs, so it need only score half the number of runs. This was soon recognized as unfair. It is much easier to score 100 runs from 25 overs than 200 from 50, because you are far less likely to run out of wickets in the shorter period. To overcome this bias towards the team batting second, the rule was changed: in the scenario described, the scorers would be asked to establish the 25 most productive overs in the first team's innings, and total up the number of runs scored in those overs alone. That would be the new target. This quickly led to anomalies: if just six overs were lost to rain, the target might not be adjusted at all! In the most notorious example, South Africa had fought their way to a position when they had the challenging, but quite feasible, task of scoring 22 runs from 13 balls; play was suspended, and when it resumed, there was time for just one ball—from which South Africa were asked to score an impossible 21 runs to win!

The main idea behind the Duckworth/Lewis (D/L) method is that, after the interruption, the relative positions of the two teams should be as little affected as possible. They proposed that not only should the target be revised to take account of how many overs were lost, it should also take account of how many *wickets* the team had left, as that also affects their chance of reaching the target. They collected data from different competitions over a number of years, fitted models, and eventually produced an answer in the form of a set of tables that scorers could consult. To see why their method is more satisfactory than previous formulae, consider a specific example. India score 250 runs in 50 overs and, after 20 overs, Pakistan are exactly on the required run rate, having scored 100 runs. Rain arrives, and when it stops, there is time for ten more overs. What should the revised target be?

If the aim is to set a *fair* target—and cricket is the epitome of fairness—we need to know how many wickets Pakistan had lost at the time play was suspended. If the opening pair were still together, Pakistan would have been very confident of winning, having 30 overs to score 151 runs to win and all wickets in hand. But if they had lost three wickets, they would have felt the game was fairly evenly poised; and if they had already lost six wickets, their task would be very difficult. Before the D/L method, the revised target would have been calculated without reference to these differences. The original pro rata method would have asked Pakistan to score 50 more runs in ten overs to tie. That should be a formality if they had lost up to three wickets; even with six wickets lost, that suddenly becomes a target they would expect to reach.

The first move in calculating the revised target under D/L is to work out what proportion of its total 'resources', which are both overs and wickets, Pakistan had used up when play had stopped. This is the information in the tables supplied to the scorers. After 20 overs in a 50-over match, a team with all wickets still intact is calculated to have 77.1% of its resources left. However, a team that has lost three wickets has only 62.3% left, and a team with six wickets down has only 36.2% left. Now look ahead to when play resumes, with time for just ten overs. In the first case, where Pakistan still has all ten wickets in hand, 34.1% of its resources remain. The interruption has taken away (77.1% − 34.1%) = 43% of its resources, so this is the equitable amount to reduce the target by. Since 43% of 250 is 107.5, remove that number of runs from India's total, revising it to 142.5. Pakistan have already scored 100 runs, so are asked to score a further 43 to win in ten overs with all wickets left. They would lose if they only scored 42, and a tie is not possible. They should be very confident—as they were before the rain. One reason why their task is now so easy is that they had been well ahead of the clock when the rain came: they were 40% towards their target, while consuming only 22.9% of their resources.

In the second case, with three wickets down, Pakistan has only 31.4% of its resources left when play resumes with ten overs to go, so (62.3% − 31.4%) = 30.9% of resources were lost to rain. Thus the target is reduced by only 30.9% of 250, i.e. 77.25 runs. India's total is adjusted to 172.75, so now Pakistan are asked to score another 73 runs (in ten overs with seven wickets standing) to win. Finely balanced—just as it was before the rain. At the time of the interruption, they had used (100% − 62.3%) = 37.7% of their resources, while getting 40% of the way to the target. One more wicket lost at the end of the 20th over would have whipped away 7.4% of their remaining resources.

Finally, the team with six wickets down has just 24.6% of its resources left when ten overs remain, so has lost (36.2% − 24.6%) = 11.6% because of the rain. The target is reduced by 11.6% of 250, 29 runs. This adjusts the Indian total to 221. Pakistan need to score another 121 runs off ten overs, with four wickets left, even to tie! But their position was already very difficult before the rain came. By losing six wickets in 20 overs, they had spent much too large a proportion of their total resources.

All the calculations here use tables based on the data available up to 1997. As more matches are played, and more accurate records kept of the running totals during the innings, the D/L tables will be refined to reflect current scoring patterns. The principles and the broad picture seem likely to remain, but the details may differ in future years. (There has been one unanticipated side-benefit. Whenever fractions of a run arise in the calculations, a match cannot end in a tie. Cricket administrators have welcomed this increased chance of a definite result!)

The D/L method can be used at any stage of the game. Suppose England bat first, expecting to face 50 overs. Play is suspended with the score 154 runs for two wickets after 30 overs. When play can resume, there is time for just 30 overs, so England's innings must be considered complete, and

Australia go in to bat. England have been deprived of what was expected to be the most productive part of their innings, the last 20 overs with eight wickets standing. They would feel justly aggrieved if Australia faced a total of 154 only, knowing from the outset they had only 30 overs to bat. But the D/L target is now rather more than 154! The tables show that England still had 54% of their resources left when their innings was halted. Now turn to Australia, who have been 'deprived' of their first 20 overs: that still leaves 77.1% of their resources, as they have all wickets intact. They have lost $(100\% - 77.1\%) = 22.9\%$ of resources. The difference between these figures is $(54\% - 22.9\%) = 31.1\%$. That is the amount of extra resources Australia have when they begin their reply, over and above what England had had. Figures over several years show that the average total in this standard of 50 over matches is 225, so England's score is increased by 31.1% of this, 69.975. England's nominal total is calculated as 154 (actual) + 69.975 (extra, because of the resources imbalance), i.e. 223.975. Australia must score 224 to win from their 30 overs. Some cricketers were bewildered by these calculations that conjured phantom runs from thin air. Politicians may claim that if their decisions are unpopular with every section of society, then their decision was probably right. In a similar fashion, if the Duckworth/Lewis calculation puzzles both teams equally, perhaps both will concede the main point, that it is *fair*!

One final example to show the supreme adaptability of the new method: suppose Kent score 200 runs in a 50-overs-per-innings match, Essex begin their innings in reply, but there are frequent interruptions for rain. There is always the chance that a downpour will bring the match to an end. One sensible rule is that so long as Essex have not lost all their wickets, or already reached the target, they must face a minimum of ten overs for a definite result to be achieved. To keep the players informed, the scorers will post a 'par' score at the end of each over; this is the score that Essex should have achieved to be smack on target. If at least ten overs have been bowled, and the match has to be abandoned before the end, Essex are declared the winners if they are ahead of par, and Kent would win if Essex were behind par. Because the par score depends both on runs to be achieved and wickets lost, the two dressing rooms will be making constant calculations. As the skies blacken, frantic signals may be sent to instruct the batsmen either to defend their wickets resolutely, or risk all on one big hit. Until the par score is reliably updated ball by ball, a team without its own lap-top computer, pre-programmed to advise on these close decisions, does not deserve to win!

Tennis

Tennis commentators will frequently refer to some points as being 'more important' than others. Pancho Gonzales, probably the best player of his time, felt that the most important point was when the score was 15–30. He argued that if the server won the next point to reach 30–30, he would very likely win his serve, but a score of 15–40 was a potential disaster. Other players and coaches argued for different points as the most crucial. The argument was effectively settled by a statistician, Carl Morris, who made a precise definition of one point being more important than another, and then presented a logical argument that identified the most crucial point.

Morris pointed out that the best approach is to compare the chances of winning a game under two scenarios: first, if the server wins the next point, and second if she loses it. The *importance* of the point is the difference between these two chances. One consequence of this idea is that the importance of any point is the same to both players—if one player loses a point, the other one wins it. As an example, we will use Morris's idea to show that Gonzales was certainly wrong: the point 30–40 is always more important than 15–30.

For, suppose the server wins the next point from either of these two scores. Then the score is deuce or 30–30, respectively. But, as every tennis player knows, these two scores are exactly equivalent: to win from either score you must either win the next two points, or share the next two points and begin again. Thus, the chance of winning from 30–30 is the same as the chance of winning from deuce.

Now suppose the server loses the next point. From 15–30, the score moves to 15–40, which is quite dangerous, but at least the server is still in with a chance. The probability she wins this game is bigger than zero. However, losing the next point from 30–40 is terminal—she has lost the game. Thus the difference between the probabilities is greater when the score is 30–40 than when it is 15–30. On any sensible measure, 30–40 is a more important point than 15–30.

Morris found there was no simple universal rule that applied to all possible comparisons. Sometimes, the relative importance of two points altered with the value of p, the server's chance of winning a point. But assume p exceeds 50%, as it does at most levels of play. At such times, Morris found that 30–40 was indeed the most important point. When p lies between 50% and 62%, the second most important point is deuce (or 30–30, of course) but when p exceeds 62%, this second most important point is at 15–40. The least important point is, not surprisingly, when the score is 40–0. The calculations that lead to these answers are not intrinsically difficult, but there are too many of them to give here.

When serving, a player has a second chance if the first serve is a fault. The usual tactics are to use a Fast first serve, and a Steady second serve if the first one is a fault. The Fast serve is more likely to win the point if it is not a fault, the Steady serve is less likely to be a fault. Are these usual tactics the only sensible ones? What is a player's best serving strategy?

His aim is to give himself the best chance, overall, of winning the point. Stephen George pointed out that the answer would come from knowing four probabilities:

(1) that a Fast serve is not a fault: call this x—perhaps 50%

(2) that a Steady serve is not a fault: call this y—maybe 90%

(3) of winning the point after a good Fast serve: call this F—80% is reasonable

(4) of winning the point after a good Steady serve: this is S—maybe 50%.

We assume that y is bigger than x, and that F is bigger than S; this sets up a tension between the two types of serve. At any point, the server has a choice of four strategies: Fast, then Steady; Fast both times; Steady both times; and Steady, then Fast. We look at each in turn, and compute the overall chance of winning the point.

Using the typical (Fast, Steady) order, the server wins the point if:

(1) the Fast serve is good, and he subsequently wins the point, or

(2) the Fast serve is a fault, the Steady serve is good, and he goes on to win the point.

For (1), the probability is $x \times F$; for (2) it is $(1 - x) \times y \times S$. Add these together to get his overall chance. Repeat this recipe for the other three possible strategies.

The first pleasing result comes from comparing the results of (Fast, Steady) and (Steady, Fast). Whatever the probabilities, the former is better. There are no circumstances (barring the element of surprise) where (Steady, Fast) should be used. But that is the only universal answer. Depending on the values of the probabilities x, y, F, and S, each of the other three strategies may be the best. The key is the ratio of $x \times F$ to $y \times S$; the former is the chance that a Fast serve is good, and the server wins the rally, while the latter is the same for a Steady serve. Call this ratio R.

Usually, R will be less than 1. But if it exceeds 1, the server should use his Fast serve both times. This is because when R exceeds 1, he has a better overall chance of winning a rally that starts with his Fast serve, even taking into account the greater likelihood that it is a fault. On fast courts in top men's tennis, this event is not at all unlikely, yet very few players, apart from Goran Ivanisevic, seem to risk a Fast second serve.

When R is just less than 1, the standard tactics of (Fast, Steady) are best. But if R is small enough—i.e. the overall chance of winning a rally that begins with a Steady serve is high enough—it is best to use a Steady serve both times. The transition point is when the ratio R is less than $1 + x - y$, and this is most unlikely to happen in professional tennis. Some sample values of the probabilities are offered above; for them $x \times F = 40\%$ and $y \times S = 45\%$, so the ratio $R = 40/45$ is less than 1, and (Fast, Steady) is best. But if the server won only 40% of points after a good Steady serve, $y \times S$ would be only 36%, R would exceed one, and the server ought to use his Fast serve every time.

Because some points are more important than others, it may be good long-term tactics to conserve effort at the less important points, in order to have more energy for the crucial ones. In Chapter 8, we saw how to translate the chance that Smith wins a single point at tennis into the chance of winning a game. That model assumed her chance of winning any point remained fixed during a game. We can adjust our calculations to allow this quantity to vary during a single game. Can she win as many games by using such a policy, while expending less energy?

Our model of a game of tennis is imperfect, and we do not expect the formulae we found to be followed with metronomic precision in real games. But if we make small changes in the model, while retaining plausibility, it is reasonable to expect the *changes* in the chances of winning a game to reflect the influence of changes in the model. Suppose that, instead of trying just as hard at every point, Smith has two levels of play. Generally, she will seek to play at the lower level, unless she is in danger of losing the game. Specifically *she plays at the lower level at the first point, and whenever she is ahead, and when she is 0–40 down; she raises her game at all other scores.* That means from deuce she will play the next point at the higher level; if she wins it, she is ahead so she reverts to lower level, but if she loses it, she continues at the higher level. Should the score reach 0–40, she expects to lose the game anyway, so does not go all out at the next point; but if she then reaches 15–40, she has new hope and switches to the higher level. The aim is to use each level of play about equally frequently.

Table 12.2 Different combinations leading to the same overall chance of winning a single game in lawn tennis.

Chance at the lower level	Chance at the higher level
58.5%	61%
57%	62%
55%	63%
53%	64%
50.5%	65%

We will compare these new tactics with the policy of playing at a constant level, and winning each point with probability 60%. The formula in the previous chapter shows that there she would expect to win 73.5% of games. With her new tactics, she wins less than 60% of the points at the lower level, but more than 60% at the higher level. Our aim is to identify combinations that still lead to her winning 73.5% of games.

The formula for winning a game is no longer as succinct as that given earlier, but it is easy enough to use. If she switches between the higher and lower levels in the way described, the combinations shown in Table 12.2 win just as many games as winning a steady 60% of points.

The table does show these new tactics are potentially useful. The *average* probability on any row is *less than* 60%. For example, if she could raise her game and win 65% of points when necessary, she could afford to play at a level that won only 50.5% of points the rest of the time. Overall, compared with a winning chance of 60% every point, it looks as though she could do just as well while (possibly) expending less energy.

Tournament formats

For any tournament—the baseball World Series, the Wimbledon tennis championships, the pool competition at the Frog and Nightgown—it could be argued that the best team or player is the one that was eventually victorious. If you hold such a view, then the question of how likely it is that the best player will win is content-free. But at the outset of the competition, someone will be favourite, and it is interesting to explore how the format chosen affects the probability that that player will win. Of course, many competitions have subsidiary requirements, beyond finding a winner: people and teams want to play a decent number of games. When

32 teams gather for the soccer World Cup Finals, it would be harsh to send any home after the first day.

There are many different formats in common use. In a simple round robin tournament, each team plays every other team once only. In soccer leagues, each pair of teams meet both home and away, but in knockout Cup competitions, there is usually a random draw each round. Tennis uses this knockout format, but the players are seeded to keep the favourites apart in the early stages. In chess or bridge tournaments, with too many contestants for all to play each other in the time available, a 'Swiss' format is used: after an initial random draw, contestants with equal, or nearly equal, scores are paired in subsequent rounds. Some croquet tournaments use a 'draw and process', with or without seeding: the draw is a knockout competition, and the process is a parallel knockout competition with the same contestants, but carefully organized to make sure that the same pairs cannot meet in the early rounds of both events. If the draw and the process have different winners, the two meet in a grand final. In all these contest formats, there is also the choice of a single match, or best of three, five or more.

David Appleton made a statistical comparison of several formats. He assumed that each player had some rating, the higher the better. When two players met, the bigger the difference in their ratings, the bigger the chance the higher rated player would win. He used the same set of ratings across several tournament formats having either eight or 16 players, and simulated ten thousand competitions in each one. In this way, differences in the tournament outcomes could be attributed to the differences in the formats. The results depended to a small extent on the number of players, and also on the precise way the rating differences translated into winning chances. But Appleton was able to reach some broad conclusions.

Suppose it is intended that the chance the favourite wins should be as large as possible. Then, of the eight formats he examined, a round robin in which each pair met twice was best. This is not surprising, as this format uses more games than any of the others. Second came the seeded draw and process. This format has two ways of helping the favourite: first, the seeding means he is spared the possibility of meeting a strong player time after time, and second he has two chances to get to a grand final. Third came a knockout competition, but where each contest was played as best of three. We noted in an earlier chapter that a best-of-three format gave a favourite a second chance, following an unexpected defeat. Close behind came the unseeded draw and process. All the above were superior to a

Table 12.3

	Anne	Bob	Cathy	Dick
Anne		A	A	D
Bob	A		B	B
Cathy	A	B		C
Dick	D	B	C	

single round robin, and the seeded knockout format. Bringing up the rear were the Swiss format, with five or six rounds, and an unseeded knockout. The Swiss and round robin formats have the disadvantage that there may be a tie for first place, whereas all the other methods give a single winner.

Several methods of resolving ties are in use. Suppose four players play a round robin tournament in a game where individual contests cannot be drawn, with the results as shown in Table 12.3. Anne and Bob each have two victories, Cathy and Dick each have one. How best to separate them? One suggestion is to take account of the individual contests. This places Anne above Bob, and Cathy above Dick. Dick may feel aggrieved, as he is placed bottom, although he beat the tournament winner!

A method often used in chess tournaments is to add up the scores of the opponents you have beaten. Anne beat Bob and Cathy, whose scores are two and one, so now Anne's score is three; similarly, Bob scores two, Cathy one, and Dick two. Anne appears to have won, but now Dick is level with Bob! We could try to separate them by repeating this adding up, but using these new scores: Anne, Bob, and Dick now get three, Cathy has two! Things appear to be getting out of hand. However, if this process is repeated just twice more, we get a definite order: Anne has eight, Bob has six, Dick has five, and Cathy three. Even better, if you repeat this process as often as you like, the scores will change, but the *order* A, B, D, C stays fixed for ever. That looks pretty fair.

Any properly organized tournament will have a specified method of separating ties. But when you are engaged in 'friendly' contests, perhaps no tie-break has been agreed in advance. If you have a battery of possible methods—I have suggested just three among the many possible—you could use a piece of nifty thinking to see which method suits your own interests best. Purely to resolve the problem, of course! No insider trading in our armoury.

CHAPTER 13

Lucky for some—miscellanea

Just as eight-ninths of an iceberg stays hidden below the surface, so only a small part of probability can be revealed in one introductory book. But there are a number of other accessible topics, collected here under this omnibus title.

Cards–whist, bridge

There are $^{52}C_{13} = 635\,013\,559\,600$ ways of selecting 13 objects from 52, so that is the total number of different bridge or whist hands that you could hold. From the work in Chapter 7, you would need to hold about a million hands for a 50% chance of being dealt a repeat, and it is most unlikely that you would notice! Effectively, every hand is a new challenge, even without the help of your 'partner', the polite name applied to the person opposite. The number of different *deals* is much larger. Whatever your hand, there are then $^{39}C_{13}$ ways of choosing the cards of your left-hand opponent, then $^{26}C_{13}$ ways to obtain your partner's hand, all of which leave some definite hand for your other opponent. Combining these, we have the 29 digit number,

$$53\,644\,737\,765\,488\,792\,839\,237\,440\,000$$

of different deals. The most sophisticated Bridge bidding system cannot hope to cater for all possible layouts.

Counting tells you how likely you are to hold certain types of hand, such as the distribution across the four suits. How often will you be dealt a void, an eight-card suit, or a 25-point hand? How often must we put up with a flat 4–3–3–3 or 4–4–3–2? Should we play to drop the Queen, or finesse?

The logical approach to distributions is in two steps, exemplified with the 4–4–3–2 question. First, fix on a specific way this can happen: four

Spades, four Hearts, three Diamonds, two Clubs. By basic counting principles, as in Appendix I, this selection can arise in $^{13}C_4 \times {}^{13}C_4 \times {}^{13}C_3 \times {}^{13}C_2 = X$, say, ways. Then ask how many ways there are of permuting the suits about. Here we can choose the two four-card suits in $^4C_2 = 6$ ways; there are then two ways of selecting one of the other two suits to have three cards, which now fixes which suit will have just two cards. So there are $6 \times 2 = 12$ ways to permute the suits, retaining a 4–4–3–2 split. The total number of Bridge hands with this shape is $12X$; earlier, we found the total number of hands possible, so the probability of a 4–4–3–2 distribution is their ratio.

Whatever distribution is considered, there are only three different answers to that second question of permuting the suits: when all four suits are of unequal length (e.g. 6–4–2–1) the answer is 24; with just two of the four suits having equal length, as in 4–4–3–2, we get to 12; and if three numbers are equal, as in 4–3–3–3, the answer is 4.

There are 39 possible distributions, ranging from the most frequent 4–4–3–2 to a one-suited hand (that you will see only when the deal has been fixed). Of these 39 possibilities, only the 13 listed in Table 13.1 have individual frequencies of more than 1%. That table also shows that you hold one of the other 26 distributions less than one time in 20. There are some curiosities: many players are surprised that the most even split, 4–3–3–3, is less frequent than four other shapes. Perhaps you do indeed hold this even distribution rather more often than the 10.5% indicated, because of poor shuffling. Table 13.1, and all the other tables, rest on the assumption that the cards are thoroughly mixed. To overcome human inadequacies in shuffling, many bridge tournaments use computers to generate hands that will fit our model closer than Uncle Fred can manage.

By collecting together the distinct possibilities in this table (and its extension to the other infrequent distributions), we can find the chance of holding very long suits, voids or singletons. There are three ways (4–4–3–2, 4–3–3–3, or 4–4–4–1) in which the longest suit has only four cards, so summing their frequencies gives the chance that no suit has more than four cards. For this, and other lengths, Table 13.2 shows what to expect. About 5% of hands contain a void, and nearly 31% have a singleton but no void. Only one hand in ten thousand is completely two-suited.

You and your partner must have some suit with at least seven cards between you, and there will be a suit in which together you have eight or more cards nearly 85% of the time. One deal in ten, the two of you have a

Table 13.1 The 13 most frequently occurring distributions of bridge hands and their probabilities.

Distribution	Probability	Cumulative prob.
4-4-3-2	0.2155	0.2155
5-3-3-2	0.1552	0.3707
5-4-3-1	0.1293	0.5000
5-4-2-2	0.1058	0.6058
4-3-3-3	0.1054	0.7111
6-3-2-2	0.0564	0.7676
6-4-2-1	0.0470	0.8146
6-3-3-1	0.0345	0.8491
5-5-2-1	0.0317	0.8808
4-4-4-1	0.0299	0.9107
7-3-2-1	0.0188	0.9296
6-4-3-0	0.0133	0.9428
5-4-4-0	0.0124	0.9552

Table 13.2 The chances that the longest suit in a given hand has specified length.

Longest suit	Probability	Cumulative prob.
4	0.3508	0.3508
5	0.4434	0.7942
6	0.1655	0.9597
7	0.0353	0.9949
8 or more	0.0051	1

ten-card fit, or better. These figures come from using the same counting principles as those leading to Table 13.1, but now we select 26 cards, not 13. The most likely distribution of your collective 26 cards is 8–7–6–5, which happens in nearly a quarter of all deals. When you hold this pattern, so do your opponents.

The democratic split of one Ace to each player happens about one deal in nine or ten, and by far the most likely event is a 2–1–1–0 spread, 58.4% of all deals. These answers arise from modifying the approach to suit distributions. For any split, first decide how many Aces each player will get; then consider each player in turn, counting how many ways there are to give him the required number of Aces, and the appropriate number of non-Aces. After 2–1–1–0, the order is 3–1–0–0, 2–2–0–0, 1–1–1–1 with 4–0–0–0 bringing up the rear, one deal in a hundred.

The most common method of hand evaluation, Milton Work's point count, awards 4, 3, 2, and 1 points to A, K, Q, and J. The total number of points available is 40, the average number is ten, but since possible holdings range from zero to 37, the distribution is not symmetric. (Extra points are usually added for long suits, or voids, but I will ignore these and look only at high card points.) The counting needed to obtain the frequencies of hands of different strength is quite complex—imagine all the different ways a hand could have ten points—but organizing a computer to make the counts is not at all hard. Table 13.3 shows the main results.

This table shows near-symmetry, around a value of about 9.5 points. The median (Appendix III) number of points is ten, which is also the most likely number, and the average. The standard deviation is just over 4. At the extremes, your hand will have three or fewer points about one deal in 20, and you will look approvingly at 20 points, or more, less than one per cent of the time. A rock-crusher of 25+ points comes your way only one deal in 2000.

After the bidding and play to the first trick, dummy is exposed; every player can now see at least 26 cards, and it is time to take stock and assess how the remaining cards are distributed. The bidding is likely to have given clues, but pure counting can be very helpful as you work out where your side can win the number of tricks it needs. Especially in no-trumps, a long suit can be the source of tricks with Sixes and Sevens, as well as Aces and Kings; what matters is how the opponents' cards are divided between their hands.

Table 13.3 Summary of the chances a bridge hand holding different numbers of points.

Points	Probability	Cumulative prob.
5 or fewer	0.1400	0.1400
6	0.0656	0.2056
7	0.0803	0.2858
8	0.0890	0.3748
9	0.0935	0.4683
10	0.0940	0.5624
11	0.0894	0.6518
12	0.0803	0.7321
13	0.0691	0.8012
14	0.0570	0.8582
15	0.0443	0.9024
16 or more	0.0976	1

Take the example of your side holding eight Clubs. The opponents' Clubs will split 5–0, 4–1, or 3–2. The latter split is likely to be favourable, but how often will it occur? Is a 3–2 split more likely than a successful finesse (50%)? Ignoring anything except the 26 cards held by you and partner, your left-hand opponent can have three Clubs in $^5C_3 = 10$ ways, and then her remaining ten cards are chosen from the 21 non-Clubs in $^{21}C_{10}$ ways. This gives $10 \times (^{21}C_{10})$ hands she can hold that contain three Clubs. As there are $^{26}C_{13}$ hands she could hold altogether, the chance she holds exactly three Clubs is the ratio $10 \times (^{21}C_{10})/(^{26}C_{13}) = 0.3391$. Her partner has the same chance, so the chance of a 3–2 split is twice this, close to 68%. All the calculations leading to Table 13.4 are made in this way.

Table 13.4 can give useful pointers at several junctions. In the play, a 3–2 split is more likely than that a single finesse will win, but a successful finesse is much more likely than a 3–3 split. You may be faced with decisions in the bidding where this table may help. Perhaps you are confident that two no-trumps will succeed; ought you to bid three?

Table 13.4 How the opposing cards are likely to split.

Cards your side holds	Opponents' cards split	Probability
7	3-3	0.355
	4-2	0.485
	5-1	0.145
	6-0	0.105
8	3-2	0.678
	4-1	0.283
	5-0	0.039
9	2-2	0.407
	3-1	0.497
	4-0	0.096
10	2-1	0.78
	3-0	0.22
11	1-1	0.52
	2-0	0.48

Your decision will rest on your assessment of the chance of taking nine tricks. In rubber bridge, the value of holding a part score is not fixed, but to simplify matters assume that the bonus points given at duplicate bridge, 50 points for a part score, and 300 for a non-vulnerable game, fairly reflect these outcomes. Set out your options thus, and note how many points you would score:

	Make eight tricks	Make nine tricks
Bid 2NT	120	150
Bid 3NT	−50	400

If the chance of making nine tricks is p, so that the chance of making eight is $1 - p$, the average score from a 2NT bid is $150p + 120(1 - p) = 120 + 30p$. Similarly, the average score from a 3NT bid is $400p - 50(1 - p) = 450p - 50$. The latter is bigger if p exceeds 17/42, about 40%. So a non-vulnerable 3NT, bid in the hope that the opponents' Hearts split 3–3 for your ninth trick, is a losing proposition—Table 13.4 shows that chance is under 40%. Vulnerable, your undertrick costs 100 points, but the game bonus rises to 500. The same considerations indicate that here it is worth pushing on if the chance of nine tricks exceeds 22/67, about 33%. That means that *vulnerable*, you show a profit, on average, if you bid 3NT hoping for a 3–3 split. 3NT bid on a finesse is expected to show a profit, vulnerable or not.

This analysis takes the average points scored per hand as the appropriate measure of success. Different considerations arise in duplicate bridge, as you have to assess what the other pairs in your seats will do.

Another problem is whether to convert a certain Small Slam into a Grand Slam that may fail. Suppose you are vulnerable with a game worth a bonus of 500, a Small Slam worth another 750, and a Grand Slam worth 1500. In no-trumps, the table is

	Make 12 tricks	Make 13 tricks
Bid Small Slam	1440	1470
Bid Grand Slam	−100	2220

Again, taking the probability of obtaining the higher number of tricks as p, the comparison is of $1440 + 30p$ with $2320p - 100$, so for the extra mile to be justified, p has to exceed 154/229, about 67%. Non-vulnerable, the arithmetic is different but this criterion is almost unchanged. So if the Grand depends on a 3–2 split in one suit (a 68% chance), go for it (just). But definitely not on a finesse. The odds have to be rather better than 2:1 on to justify one more for the road at the six-level.

Missing the Queen in a suit where your side has the majority of cards, what are the respective chances of picking it up if you play for the drop, or cash one high card and then finesse? When you hold nine cards, split 5–4, the drop play succeeds if the Queen is singleton, or the opponents' cards split 2–2, a total chance of 53.1%. There is a further chance—you are able to finesse the Queen after you discover the split is 4–0, which takes the total to nearly 58%. The finesse play wins when you can pick the Queen up despite a 4–0 split, when the Queen is singleton, or when it is otherwise favourably placed: these chances add up to just over 56%. There is little to choose between the two plans—but maybe one opponent is trying to look nonchalant? With four cards in each hand, you cannot pick the Queen up if the split is 5–0. Playing for the drop succeeds nearly 33% of the time, the finesse brings the bacon home with probability 51%. When you have only seven cards, the finesse is the better bet even more comfortably.

You may be able to postpone locating this Queen until several tricks have been played; perhaps an opponent will be considerate enough to discard in the same suit. But if no such carelessness occurs, the reduced number of cards will still change the calculations. Suppose seven tricks have been played, you and partner hold nine Spades without the Queen, and no Spades have yet been played. The opponents now hold 12 cards, four Spades and eight non-Spades; the chances of 2–2, 3–1, and 4–0 splits have changed to 45.5%, 48.5%, and 6%. That makes your chances of picking up the Queen at over 60% playing for the drop, 56% relying on the finesse. If you were inclined to go for the drop at trick 1, that play is even more favoured now; but the overall chances do not change greatly until nearly all the cards have been played.

Shuffling

There are two basic types of shuffle, the riffle shuffle and Monge's shuffle. In a perfect riffle shuffle, the pack is split into two halves, and the cards interleaved one at a time. There is a difference, according to whether or not this riffle shuffle leaves the previous top card in its place, or moves it to the second place. If the top card remains in place, so does the bottom one, and careful counting shows that the cards in 18th and 35th places swap round. The remaining 48 cards divide into six blocks of eight cards, and within each block the cards cycle around. Thus, after eight perfect riffle shuffles, *every card is back in its original place*! Card players must not be too expert at a riffle shuffle, or their objective of mixing the cards up can be thwarted by their own proficiency. If the two half-packs are swapped around, so that the card at the top now moves to second place, it takes 52 perfect riffle shuffles to regain the start order.

Monge's shuffle is usually accomplished by slotting the cards from one hand to the other; card 1 is dropped, card 2 placed above it, card 3 below it, card 4 above card 2, and so on. Accomplished perfectly, this leaves card 18 in place, interchanges 11 and 32, and breaks the other cards into cycles of lengths 3, 4, 6, and three cycles of length 12; after 12 perfect Monge's shuffles, all the cards return home.

Fortunately, few players are dextrous enough to make one perfect riffle shuffle, let alone eight in succession. Richard Epstein reported that shuffles take between 0.7 seconds (beginner) and 0.5 seconds (virtuoso), with beginners tending to interleave clusters of cards near the top and bottom, getting singles or pairs in the middle. Professional dealers create single-card interlacings about eight times as often as a pair slips in, and usually avoid a group of three cards. How many shuffles are enough to mix the cards?

In recent years, this problem has received attention from serious probabilists, motivated by the increasing use of computers to simulate processes that are too complex for mathematics to deal with entirely. Suppose a new supermarket is planned: the operation of alternative layouts of the delivery bays, the interior, the check-outs, etc. can be simulated on a computer much more cheaply than building first and amending later. But any computer simulation of a day's operation will allocate some initial status to all variables—number of customers, what items are short, when deliveries will come—and it is important that these initial assumptions do not have a disproportionate influence on the predictions. After a while, the operation of the supermarket will have settled down to some statistical equilibrium— but after how long? That is the equivalent of asking how many shuffles produce a random deck.

The expert in this area is the American statistician, Persi Diaconis, whose other talents include stage magic, and the ability to accomplish perfect riffle shuffles. Along with colleagues, he looked at the mathematics behind different ways of shuffling, and found one absolutely fascinating result: whatever the nature of the shuffle, the deck remains conspicuously similar to its original order for a period, but then, over a relatively brief time, becomes effectively randomly ordered. The time of this jump, from order to chaos, depends on the shuffling method, but it can be calculated. For the simple 'shuffle' in which the top card is placed at random amongst the other 51 repeatedly, you need about 240 movements. If you use standard imperfect riffle shuffles, eight or nine will usually suffice. Five shuffles are definitely too few.

Because of the cyclic nature of perfect riffle shuffles, an expert shuffler may not randomize the cards as well as a clumsy amateur. Richard Epstein likens a genuinely random shuffle to the scattering of cards in high winds, retrieved by a blindfolded inebriate. You need not try quite so hard.

Dividing the prize

Harry and Stanley are playing cards. The first to win six games takes the prize, but Stanley, a brain surgeon, is called to an urgent operation when leading 5–4. What is an equitable division of the prize?

Like Chevalier de Méré's query over throwing pairs of dice to get a double Six (Chapter 5), this puzzle has an honoured place in the history of probability. It was the correspondence between Pascal and Fermat in 1654 as they exchanged ideas on this 'problem of points' that is usually cited as the first systematic mathematical treatment of a probability problem. They had different approaches, but reached identical answers. Essentially, their method was to ask how many points short of the target the contestants were: and then to suppose that, had the match continued, each player was equally likely to score the next point, until the end. With this model, the probability that either player would win can be found, and a fair division of the prize is to allocate a share to each player, in proportion to his chance of winning.

In the initial example, Stanley needs one point and Harry needs two. Harry can only win if he scores the next two points, which will have probability $(1/2) \times (1/2) = 1/4$. So Harry gets 1/4 of the prize, and Stanley the other 3/4. If they had needed equal numbers of points, the prize would be split 50–50. Table 13.5 shows the probabilities each player wins, and hence how to split the prize, when Harry and Stanley are near the target.

Table 13.5 Stanley's chances of winning, when he is x points short and Harry is y.

Stanley/Harry	$y=1$	2	3	4
$x=1$	1/2	3/4	7/8	15/16
2	1/4	1/2	11/16	13/16
3	1/8	5/16	1/2	21/32
4	1/16	3/16	11/32	1/2

Extending this table is easy. To find what would happen if Harry were two points short, and Stanley were five short, argue that after one more game, it is equally likely that Harry is one short and Stanley five short, or that Harry is two short and Stanley is four short. In the first case, Stanley's winning chance is plainly 1/32, in the second case the table shows his chance is 3/16. So his chance is the average of these, 7/64. Any split you like can be found in this fashion, by considering one game at a time.

Vowel or consonant?

Sam selects a page at random in a novel, and then a letter at random on that page. Polly seeks to guess whether that letter is a vowel or a consonant. What tactics should she use, and how likely is she to be successful?

That preceding paragraph has 66 vowels (counting 'y' as a vowel in 'Polly' and 'likely') and 105 consonants. This count confirms the instinctive feeling that consonants occur more often. To maximize her expected number of successes, Polly should guess 'consonant' every time. Her success rate is the frequency of consonants in the novel.

Sam decides to be more helpful. He still selects a letter at random in the same way, but now tells Polly whether the *preceding* letter (in the same word, or not) is a vowel. Is this information useful? If it is, how should Polly amend her strategy?

Using that same first paragraph to give guidance, a vowel is followed by a consonant 59 times, and by another vowel only seven times. If the preceding letter is a vowel, the guess 'consonant' has a very high chance of being correct. Similarly, a consonant is followed by a vowel 59 times, by another consonant 45 times. Provided such a small sample does not mislead, it seems the extra information is useful, in that Polly should now guess 'vowel' if the preceding letter is a consonant.

This innocent problem was an early example of the study of what are now called Markov chains. In 1913, the Russian probabilist A. A. Markov gave the results of his analysis of 20 000 letters from Pushkin's tone poem *Eugene Onegin*. He found that 43% of the letters were vowels, 57% consonants, so the strategy of guessing consonant in the absence of other information was plainly best. Looking at the change between consecutive letters, a vowel was followed by a consonant 87.2% of the time, a consonant was followed by a vowel 66.3% of the time. So the strategy of guessing the opposite of the previous letter would maximize the success rate.

But the real importance of Markov's study was the next stage. What extra help accrues if the *two* previous letters are given? Markov found that there was no appreciable help at all. To be told the two previous letters were 'cv' or 'vv' was no more useful than that the last letter was 'v'; and 'cc' or 'vc' were no more useful than just 'c'. In the same fashion, knowing three or more previous letters was no real extra help to the guesser.

Reducing Pushkin's prose to a machine-gun 'cvccvcvvcccvcv…', indistinguishable from Mills and Boon, may be literary vandalism, but Markov chains have proved of immense value. They have been applied successfully to models in many fields, including physics, biology, social science, operations research, economics, epidemiology, and sport. Their usefulness stems from the fact that they are just one step more complex than independent trials.

The assumption of independence is often convenient, as the subsequent analysis is simpler, but leaves a suspicion that the model is unsatisfactory. In Chapter 8, for example, we took it that the results of successive frames of snooker, or points within a game of tennis, were independent. But perhaps, having just lost a frame or a point, a player is *more* likely to win the next one, on two grounds: she tries harder to catch up, or her opponent pays less attention after a success. Alternatively 'success breeds success', and a player who has just won a point may have an increased chance of winning the next. Just as with vowels and consonants, what happens next may not be independent of the most recent outcome; but knowing that most recent outcome, results previous to that may often be effectively disregarded.

That is the central notion of a Markov chain. There is a history of outcomes (not necessarily restricted to just two alternatives) which we hope to use in predicting the next outcome. The sequence will be a Markov chain if, given this history, we can disregard all but the most recent result. In such a case, the influence of one outcome on the next can be summarized in a small table. Markov's study found:

	Vowel	Consonant
Vowel	0.128	0.872
Consonant	0.663	0.337

The most recent outcome is the label of each row; the entries along a row are the respective probabilities of the next outcome, so the entries on each row will always sum to a total of one.

Another example is a 27-year study of the December to February rainy season in Tel Aviv, covering 2437 days. Classifying each day simply as Dry or Wet, it was found that knowledge of the entire weather pattern up to today gave no better predictions for tomorrow than knowing just today's weather. Given today, yesterday can be forgotten. The data showed:

	Dry	Wet
Dry	0.750	0.250
Wet	0.338	0.662

If today is Dry, the chance tomorrow is Dry is 75%; but if today is Wet, the chance it is Dry tomorrow is down to 33.8%.

These tables ('transition matrices' is the jargon) can now be used to make predictions for two, three, or more steps ahead. Take Markov's original data, and suppose the present letter is a vowel. What is the chance the next-but-one letter is also a vowel? To answer this, note there are just two patterns that give the result we seek, namely v-v-v and v-c-v; the first has probability 0.128×0.128, the second has probability 0.878×0.663, and these sum to make 0.595. Thus 59.5% of the time that the present letter is a vowel, the next-but-one is also a vowel, which means that the next-but-one is a consonant about 40.5% of the time. We could check that answer by directly calculating the probabilities of v-v-c and v-c-c; and you should do so.

Three or more letters ahead, the computations are longer but the idea is the same. List all the ways of reaching the target from the initial point, calculate each probability, and add them up. To find the chance of a vowel three steps hence, given a vowel now, we must examine v-v-v-v, v-v-c-v, v-c-v-v, and v-c-c-v. Arithmetic leads to an answer of 0.345. If the current letter is a consonant, a parallel argument shows that the chance of a vowel three steps ahead rises to 0.498. The longer ahead you look, the less relevant is the current information. Looking five steps ahead, the chances of a vowel are 41% or 45%, depending on the present status; nine steps ahead, the answer is 43%, whatever the present letter. That, of course, is the overall frequency of vowels.

This is not the place to develop Markov chains any further. There are efficient ways to calculate the probabilities that relate to several steps ahead, and to find long-term averages. Because of the usefulness of Markov's ideas, splendid textbooks describe how to work with these chains in ample detail, with copious examples. My purpose is to convince you that, although there may be deficiencies in many of the simple models we have used, there is life beyond those models: but the detail of remedying the deficiencies sometimes gets rather heavy.

Benford's law

Select a fairly arbitrary piece of real data. It might be the lengths of various rivers, the individual payments received in a month by a mortgage lender, the populations of a collection of towns, or similar. The only restriction is that it should be neither systematic, nor deliberately random: no lists of telephone numbers, or of prime numbers, or the output from a computer's random number generator. Whatever the source of the data, focus on the *first significant digit*. Whether the data value is 81, 8.1, 0.81, or 0.081, that digit is 8. If Mary has such a set of real data, and selects one value in it at random, what is John's chance of guessing that first digit correctly? If Mary offered a £10 prize, how much should John pay to play the game?

There are only nine possible answers to Mary's question, as zero is never the first significant digit. If John makes guesses at random, he has one chance in nine of being correct. An entry fee of £1.11p is pretty fair. But even when Mary does not state the source of her data, I would happily pay twice that sum, and expect to be in profit, if Mary does act as described. My policy would be boring, but successful. I consistently guess that the first significant digit is 1.

The reason for my policy, and my confidence in its success, is that when first digits are chosen in this fashion, the answer 1 appears more than 30% of the time. Plainly, if that claim is correct, my average payout at each play is over £3, so I will expect to make a profit at any lower entry fee. The claim may seem far-fetched: it appears to assert that more than 30% of numbers begin with a 1. However, if you read the description carefully, the assertion is that more than 30% of the numbers *that you come across* in these arbitrary data sets begin with the digit 1.

This claim is known as Benford's law, in honour of Frank Benford who wrote about 'the law of anomalous numbers' in 1938. He had noticed that log tables (readers aged under 25 years should consult their parents) were dirtier at the beginning than at the end. He attributed this to users needing to look up the logarithms of numbers whose first digits were 1 or 2 rather more often than numbers that began with 8 or 9. His observation was not new. Simon Newcombe had written about the phenomenon in 1881, but

Benford's rediscovery sparked enough interest for later writers to cite Benford as the originator.

The full version of Benford's law, as applied to numbers written in decimal notation, is that the chance that the first significant digit is d is $\log_{10}((d+1)/d)$. For initial digit $d = 1$, this value is $\log_{10}(2) = 0.3010$, hence the claim 'over 30%'. The law says that the chance the first significant digit is 2 is $\log_{10}(3/2) = 0.1761$, and so on. See Table 13.6, which also gives support to the law by including real data from different sources.

Benford's own data on the second row of Table 13.6 were an amalgamation of figures from 20 different sources, covering over 20 000 actual values. The sources were diverse: some were from tables of atomic weights, others were the house numbers of famous scientists, some were taken from mathematical tables of reciprocals. Individually there were considerable differences from the proportions expected according to his law, but the large quantities of data swamped these deviations. Collectively, the similarities of the entries in the table for 'Benford's law' and 'Benford's data' are very striking. The data labelled '100 Towns' come from 100 consecutive entries in the 1980/81 *AA Handbook*, which lists towns alphabetically, and gives their population sizes. The 'Electricity' data come from the consumption figures of 1243 customers in one month, from Honiara in the Solomon Islands. These data were sent to Ralph Raimi by the manager of the company, after Raimi wrote about Benford's law in a scientific magazine.

Despite any initial implausibility, the data do support Benford's law, so we should seek an explanation. We have to acknowledge that the vague formulation 'a fairly arbitrary set of data' used at the beginning would need some more accurate specification if we wanted a mathematical proof.

We seek to apply the law to data of at least two different types. For populations of towns, there is a natural counting scale: one person is always just one person. But when measuring lengths of rivers, there is a

Table 13.6 The probabilities of the first significant digits, according to Benford's law, and three examples. See the text.

Digit	1	2	3	4	5	6	7	8	9
Benford's law (%)	30.1	17.6	12.5	9.7	7.9	6.7	5.8	5.1	4.6
Benford's data (%)	30.6	16.7	11.6	8.7	8.5	6.4	5.7	5.0	5.7
100 towns	28	20	13	11	5	10	9	3	1
Electricity (%)	30.6	18.5	12.4	9.4	8.0	6.4	5.1	4.9	4.7

completely arbitrary scale being used: it might be miles, inches, kilometres, cubits, leagues, yards—anything. And yet Benford's law, if it holds, should apply equally well to all these scales; if our data are all multiplied by some constant, be it 2.54 to go between inches and centimetres, or 3 to convert yards to feet, the proportion of all the data with a given initial digit should remain the same. Mathematicians have shown that the only distribution of frequencies of these digits that remains unchanged when such data are multiplied by any fixed scale factor is this logarithmic list. That is a substantial step, but not the whole way. *If* there is some distribution of frequencies that stays the same whatever scale the data are measured in, the data conform to Benford's law. But that is still a big if. Why should any such law exist?

Try the experiment of starting with some initial number, and repeatedly doubling it, noting the first digit each time. Thus beginning with 6, the doubling leads to 12, 24, 48, 96, 192, ... , and the initial digits are 6, 1, 2, 4, 9, 1, ... Continue this for some time—a hundred doublings don't take long with a pocket calculator—and see the results. You should find that the frequencies of the initial digits 1 to 9 are 30, 18, 12, 10, 8, 7, 5, 5, 5—very similar to those for Benford's law shown in Table 13.6. Doubling each time is just one example; you get a similar result if you multiply by other numbers, such as 1.5, π, 4.2, etc. (but not if you multiply by 10, of course!)

What this experiment suggests is that if your data have some wide spread ('a fairly arbitrary set of data'), as you dip into positions across that spread, you come across numbers beginning with 1 much more often than numbers beginning with 9. Try the '50% either side test'. For a range of ball-park numbers, go 50% either side, and look at the respective proportions of initial 1s and 9s. If you work your way through the list 100, 200, ... , 900 as middle numbers, and go 50% either side of each, (i.e. 50 to 150, then 100 to 300, etc.) the 1s win hands down, overall. Working from 1000 to 9000, or higher ranges, would essentially repeat this exercise.

None of this is a 'proof' of Benford's law, since any such proof would have to be precise about the types of data to which it should apply. Benford's own data—assorted additions from diverse sources—would have been thrown out of kilter if the data had also included the ages, in years, of all adult men currently held in gaol on the Isle of Wight (not enough aged exactly 18, 19, or between 100 and 199 years). There are mathematical arguments that make it seem reasonable that data covering a wide range of values should follow this law, but they rest on assumptions that real data may not satisfy. If a 'village' means a settlement whose population is between 200 and 999, Benford's law would be manifestly false for data on village populations. But if Mary plays the guessing game fairly, as described, I would take my chances, and constantly offer '1' as my prediction.

An alternative version, if Mary is playing the same role, is for John to win when the first significant digit is 1, 2, or 3 and for Mary to win if it is 4 or more. Perhaps many prospective 'Marys' would be willing to play this game at even money? However, for data that follow the law, Table 13.6 shows John can expect to win just over 60% of all bets, a comfortable margin.

Hal Varian made the interesting suggestion that Benford's law could be used to detect honesty in the reporting of socio-economic data, such as that used in public planning decisions. Provided real data follows this law, if careless planners rounded their figures and then used these less accurate figures as a basis of their decisions, that could be detected: simply compare the frequencies of the first digits of the *reported* figures with those expected from Benford's law.

An accountant, Mark Nigrini, has developed this idea further. He pointed out a possible way to detect fraud. Accounting data are a complete mix of figures of different orders of magnitude—a few lunches here, a new car there, refurbishing the head office, etc. By shifting the decimal point if necessary, all these numbers can be written so that they appear to be between one and ten. £452.50 becomes 4.525, £8.94 remains as 8.94, 26p becomes 2.6, and so on. The figures are now useless for accounts, but this transformation has highlighted the first significant digit, hence the connection with Benford's law.

We expect more of these figures to begin with 1 than with 2; more with 2 than with 3, and so on. Nigrini suggested that you now add up all the numbers that have the same first significant digit. Although those that begin with the digit 1 are smallest, we expect to have a lot of them; not very many that begin with 9, but they are quite big. Remarkably, it turns out that these nine sets of additions should all lead to approximately the same total (if the figures obey Benford's law)! One trick that might appeal to fraudsters is to attempt to hide some transaction for £150, by splitting it into two items, of £80 and £70, hoping that less scrutiny will be made of sums under the cut-off point of £100. Any such action would distort the frequencies of the initial digits, and could show up via anomalies in these 'Nigrini sums'.

Guess the other colour

Take three cards of identical size and shape. Colour one of them red on both sides, the second one black on both sides, and make the last one red on one side, black on the other. Jumble the cards in a hat, lift one of them out in such a way that one side only is seen, and place it on a table so that all can see this exposed side. Suppose it is red. The object is to guess the colour on the *other* side. What is your guess, and how often will you be right?

Warren Weaver, whose work on the theory of communication with Claude Shannon has ensured his place in science, describes this problem in his book *Lady Luck*. Like our 'find the lady' game of Chapter 9, the central notion here is how to deal with incomplete information. Spot the flaw in the following.

> Since we can see a red side, the card cannot be black-black. There are two other cards, equally likely since the card was selected at random. One of these cards has red on the other side, the other card has black, so red and black are equally likely. I'll make an even money bet on red—that will be a fair bet.

Anyone believing that, and accepting the bet, is on to a loser. The concealed side is *twice* as likely to be red as black, so the game has a strong bias in favour of the bettor on red. Weaver claims that, as a graduate student just after the First World War, he made profits out of his fellow students, treating his gains as fair payment for teaching them the ideas of probability.

There are several ways to see why red is twice as likely as black, when red is the exposed side. Here are two different approaches, both valid:

(1) Originally, there are three black sides and three red sides, all equally likely to be shown. If all we know is that a red side is exposed, any one of the three red sides is equally likely. But if we select one of these three red sides at random, and look at the *other* side, on two occasions the other side is red, and only once is it black. So red is favoured two to one.

(2) The person who bets on red when the red card is showing will presumably bet on black if the black card is exposed. In other words, this bettor is betting that the two sides show the same colour. But two of the three cards do have two sides of the same colour, and only one card has different colours. Since the card is chosen at random, betting on red when red shows and black when black shows will win two times in three.

The flaw in the set-up argument first offered lies in the assertion in the second sentence that the two cards are equally likely. Sure, all the cards are equally likely when no information is available. But to say 'I have selected one side of this card at random and it is red' changes matters. As the arguments A and B show, it is now twice as likely that the card with the same colour on each side had been chosen.

At least two Aces?

Only 26% of Bridge hands have two or more Aces. If Laura offers to deal a Bridge hand at random, and pay out at even money on a bet that it has at least two Aces, you should decline her offer. It would be a poor bet. But suppose Laura repeatedly deals out a single hand from a well-shuffled deck until she is able to say truthfully that 'This hand contains an Ace'.

Would you then be prepared to bet on at least two Aces, at even money? Or suppose Laura similarly dealt herself hands until she could truthfully announce 'This hand contains the Ace of Spades'.

Would that be a different situation? Is the second statement essentially equivalent to the first, or does it make holding at least two Aces either more likely, or less likely? Does either statement make betting on at least two Aces a good proposition, at even money?

Answers to these questions will come from an exercise in counting. But a full deck of 52 cards gives rise to far too many possibilities for easy assessment. Suppose we reduce the entire pack to just four cards, the Aces and Kings of Spades and Hearts, and let a 'hand' have just two cards. This simpler scene may give a pointer to the more complex one. With no information at all, there are just six hands possible, all equally likely. That makes the chance of 'two Aces' one in six. To be told that the hand contains an Ace merely eliminates the hand with a pair of Kings, leaving five hands. But to be told it contains the Ace of Spades knocks out three hands, leaving only three possible. In this case, the two statements lead to *different* estimates, one in five and one in three, of the chance the hand has two Aces. We should not be astonished if something similar holds with a full deck.

Before we do exact counting, it is usually a good idea to try to get a rough estimate of the answer. If our counting then gives a very different result, we ought to check why the discrepancy has arisen. It is easy to overlook something.

Think of a bridge hand in general. On average, it will have one Ace, and it seems plausible that 'one Ace' is the most likely outcome. Since the average is one Ace, hands with no Aces must arise more often than hands with two or more Aces. Thus, in order of decreasing probability, we expect to have one, zero, two, three, or four Aces, with three or four Aces fairly rare. This ordering suggests there will be a *lot* more hands with one Ace than with two Aces—perhaps twice as many. If this is true, then when Ace-less hands are eliminated, the chance of at least two Aces is about one in three.

For the second statement, our hand consists of the Ace of Spades and 12 other cards, selected at random from the remaining 51. Our hand will have two or more Aces whenever it has at least one more Ace. There are three Aces in these 51 cards, so any card dealt has chance 3/51 of being an Ace. With 12 cards, the *average* number of Aces will be $12 \times (3/51)$, about 2/3. The actual number will be zero, one, two, or three, and most of the time it will be zero or one, and to make the average 2/3, one Ace will be more likely than none. The chance of at least one more Ace seems to be between 1/2 and 2/3.

Now for the full arithmetic. Consider the information that the hand has at least one Ace. We counted how many possible Bridge hands there are altogether. Using the same methods, count how many have no Ace, and how many have exactly one Ace. By subtraction find the numbers with either at least two Aces, or at least one Ace. The ratio of these numbers is then the proportion of hands with at least two Aces among hands with at least one. After the arithmetic, the answer turns out as about 37%, plainly compatible with the rough estimate of one in three. If Laura tells us the hand has at least one Ace, we should bet *against* it having two or more.

If we are told the hand has the Spade Ace, the counting is easier. We just look at how the other 12 cards can split between Aces and non-Aces. There are $^{51}C_{12}$ ways to choose the 12 cards, and only $^{48}C_{12}$ ways of selecting all of them from the non-Aces, thereby giving an Ace-less hand. The ratio of these numbers is the chance of no Aces; from which the chance of at least one Ace comes from trick 1 (Appendix II) in probability. It is about 56%—in the middle of our guess that it would be in the range 1/2 to 2/3. This time, the bet that there are at least two Aces is favoured.

Before these arguments are shown in detail, many people will believe the two statements contain equivalent information. 'After all, if it does contain an Ace, the Spade Ace has just the same chance as any other' is plainly correct, but not relevant to the problem. The two statements eliminate different numbers of hands, as the example with a deck of just four cards shows.

Even very numerate people get the wrong answer to the original problem, because of the difficulty of assimilating the relevant information. Suppose the problem had been: 'This number is chosen at random between zero and 100. Guess whether it exceeds 60'. Then it is easy to use extra information such as 'The number is at least 40'. But with the deck of cards, either statement about Aces eliminates some of the possibilities, *but not in a simple fashion*. We cannot list all the possible hands of 13 cards as conveniently as the numbers between zero and 100.

This is the third challenge, after 'find the lady' in Chapter 9 and 'guess the other colour' above where a heated argument about the correct analysis might break out. I hope I have been careful enough in specifying my models so that the analyses I have offered are valid. But to illustrate one place where great men have disagreed, consider the reduction of the deck to four cards described above. This idea came from Martin Gardner, who contributed an entertaining mathematical column to the *Scientific American* for many years. Warren Weaver argued that Gardner's answers of one in five, and one in three, for a reduced pack were unsound. He saw no reason to believe that, if there are six equally likely possibilities, and a statement simply eliminates some of these, then those left are all equally likely. He claims that there is a parallel with the 'guess the other colour' problem.

I am happy with Gardner's analogy. There is a major difference between the two problems. With the colours, seeing a red side does rather more than simply eliminate the black-black card. All three cards were equally likely before the red side was revealed; all three *red sides* remain equally likely after the red side is revealed—but the two cards not eliminated are not equally likely, as one has twice as many red sides as the other. But with the deck of cards, the information given either does or does not eliminate a given hand, period.

I do agree with Weaver that there is more to the problem than has been said so far. We have taken the statements 'This hand contains an Ace', and 'This hand contains the Ace of Spades', and decided what possible hands are eliminated by them. As specified above, Laura was under severe constraints about what she could say. But suppose she simply dealt out one Bridge hand, and could choose from among a variety of statements:

- this hand has an Ace;
- this hand has the Ace of Spades;

- this hand has no Clubs;
- this hand has some black cards;
- this hand has at least five cards higher than a Nine

in whatever fashion she wished. Then the analysis would be completely different. We would have to assess the relative likelihood of which of two or more valid statements would be chosen, as well as how likely each is to be true in the first case. This would be horribly complicated.

I suspect that this is the root of Weaver's suspicions about Gardner's analysis—he had a different model of what the speaker might say. In Chapter 1, I pointed out the personal nature of probability. Given any 'experiment', we need to construct a model, and work out our probabilities from the model in a consistent manner. But the probabilistic model is only useful in as much as it reflects the real experiment. Two people may have different amounts of information, so their models can lead to sensible, but different answers.

Which player is favoured?

When a game is set up, it should be possible to decide whether it is equally fair to both sides, or whether one has an advantage. But it is not always clear what one side 'having the advantage' actually means. Thomas Cover explored the difficulties in this notion by presenting a series of examples; they show how the length of a contest can be crucial.

Suppose Anthony and Cleopatra play a series of games. Anthony automatically scores two points each time, while Cleopatra will score either three points or zero. Her chance of scoring three points, in any game, is fixed at $p = 0.618$. It is the accumulated score that determines the winner, not how many single contests each wins (contrast Chapter 8). Suppose the whole contest consists of just one game. Then Anthony gets two points, so Cleopatra's chance of winning is her chance of scoring three points, and this is p, nearly 62%. But if the contest is decided as the total score over *two* independent games, Anthony will be favoured. He will have four points, and will win unless Cleopatra scores her maximum each time, and the chance of this is p^2, about 38%. The roles are entirely reversed in the change from one game to two. And Anthony is more likely to win on any longer contest (ties excluded).

Change this a little. Anthony has one fruit machine, Cleopatra has a different one. They play their machines simultaneously, and compare their total scores after a given number of games. Cover constructed a splendid example to show how much might depend on the number of games played. He supposed Anthony gets 500 points each time, and Cleopatra gets 501 points with probability 99%, and zero the rest of the time. After just one game, she plainly has a 99% chance of being ahead. But after 500 games, Anthony will have 25 000 points, and if Cleopatra has scored even a single zero, her maximum possible score is 24 999. So her only chance of being ahead at that stage is to have scored 501 every time, which occurs with probability $(0.99)^{500}$, about 0.0066, considerably less than 1%. The turn-around from 99% in a contest of one game to under 1% in a much longer series is dramatic. It is clear that with larger numbers we could adapt this example to get even more extreme results.

Cover went on to use great ingenuity to construct sets of rewards in a single game that ensured that Cleopatra's winning chance would oscillate wildly, as more and more independent repetitions of this game were played. He was able to arrange it so that initially she was nearly certain to be ahead, but some time later her chance dropped to near zero. A few games later, the computations showed that her chance had risen to be closer to certainty than before, followed by a drop to become even closer to zero. This pattern can continue for ever: now her chance of being ahead is tantalizingly close to certainty, and then it sinks to the verge of being zero, before soaring even closer to one, and so on. In this contest, the choice of who is favoured depends entirely on how many games are played. After (say) 500, 1000, or 3000 contests, Cleopatra could be the overwhelming favourite, but after 800 or 2000 contests, her chance would be near zero. Constructing this example was not easy, but its existence does show what is possible, and cautions us against facile generalities like 'longer games favour the stronger player'.

The problem with that statement is that the notion of which is the stronger player is bound up with how long the contest lasts. The phrase 'stronger player' is not meaningful until the length of the contest is known. Aesop, of course, knew this. Is it the hare or the tortoise who is the stronger player?

Counting

Models of experiments that use packs of cards, collections of Lottery numbers, roulette wheels and the like commonly assume that all the distinct outcomes are equally likely. The probability of any outcome is then found through a well-organized piece of counting. One central principle solves most counting problems:

> *if there are M ways to do one thing, and then N ways to do another, there are M × N ways to do both, in that order.*

For example, suppose we want to know in how many different ways three objects, {A, B, C} can be listed. One way to solve it is just to list the ways as

ABC   ACB   BAC   BCA   CAB   CBA

which are plainly all different, and exhaust the possibilities. So there are six ways. But to answer the same question with the five objects {A, B, C, D, E}, using the same method, would be ridiculous. Instead, use the central principle to argue as follows:

'There are $M = 5$ ways of choosing the first object. After that choice, there are $N = 4$ ways to select the second, so there are $5 \times 4 = 20$ ways to choose the first two objects, in order. Having chosen two objects, there are just three ways to select the third, so that gives $20 \times 3 = 60$ ways for the first three objects. Now, that leaves just two choices for the fourth object, leading to $60 \times 2 = 120$ ways for the first four objects. The last object goes inevitably in the fifth place, just one way to place it, making $120 \times 1 = 120$ ways to place all five in order. Without writing any of them down, we know there are 120 ways.'

Note how this answer was reached. The calculation that gave the number of ways to order five objects was

$$5 \times 4 \times 3 \times 2 \times 1 = 120.$$

There is a convenient shorthand for the calculation $5 \times 4 \times 3 \times 2 \times 1$; it is written as 5!, pronounced 'five factorial'. With just three objects, there are $3! = 3 \times 2 \times 1 = 6$ ways to list them in order, as we saw. Summarizing:

Result 1

For any collection of K different objects, there are 'K factorial', i.e.

$$K! = K \times (K-1) \times (K-2) \times \ldots \times 3 \times 2 \times 1$$

different ways of listing them in order. Table AI.1 gives the first few values.

As the number of objects increases, the number of ways of listing them in order increases very rapidly. With a deck of playing cards, there are 52! different results from shuffling—this number has 68 digits!

Table AI.1 The first few values of $K!$

Number of objects, K	1	2	3	4	5	6	7	8
Number of orders, $K!$	1	2	6	24	120	720	5040	40 320

As another example, suppose we wish to select two from a list of six objects. We can select the first in $M = 6$ ways, and then choose the second in $N = 5$ ways, giving $6 \times 5 = 30$ ways to choose both. But this calculation includes any choice such as {A, B} *twice*, once in the order AB and once as BA. Our value of 30 is thus *double* the answer we seek, so there are 15 different pairs of objects from six, if we disregard the order in which they were selected.

The same argument will work for a collection of any size, K. There are K ways to choose the first, then $(K-1)$ to choose the second, so $K \times (K-1)$ ways of selecting both. This answer counts every pair twice, so

Result 2

There are $K \times (K-1)/2$ ways of choosing *two* objects from a collection of size K, paying no regard to their order.

Counting how many ways we can choose three objects from a collection of size K is organized in the same way. First, count how many ways if we did pay attention to order; then assess how often each choice of three has been made. For example, how many committees of size three can be chosen from seven eligible members? Taking account of the order of selection, the previous methods show there are $7 \times 6 \times 5 = 210$ ways. But any choice such as {Alice, Bill, Chris} will have been made $3 \times 2 \times 1 = 6$ times. The first answer of 210 is six times too big, so the actual number of different committees is 210/6 = 35.

Spelling out the calculation, the answer came from

$$\frac{7 \times 6 \times 5}{3 \times 2 \times 1} = 35.$$

Plainly, with ten eligible people, we could choose

$$\frac{10 \times 9 \times 8}{3 \times 2 \times 1} = 120$$

committees of size three; and selecting five from a group of 12 can be done in

$$\frac{12 \times 11 \times 10 \times 9 \times 8}{5 \times 4 \times 3 \times 2 \times 1} = 792$$

ways. Similar calculations arise so often that another piece of shorthand is useful. Write $^{12}C_5$ (and pronounce it '12 choose five') for that last calculation. The earlier ones showed that $^{10}C_3 = 120$, $^{7}C_3 = 35$, and Result 2 could be written as $^{K}C_2 = K \times (K-1)/2$. In the same manner,

$$^{K}C_3 = \frac{K \times (K-1) \times (K-2)}{3 \times 2 \times 1}, \quad ^{K}C_4 = \frac{K \times (K-1) \times (K-2) \times (K-3)}{4 \times 3 \times 2 \times 1}$$

and so on.

Result 3
The number of ways of selecting r objects from a collection of K objects, paying no attention to the order, is written $^{K}C_r$, pronounced 'K choose r'. Its value is found as illustrated above.

The really important thing is that *some such formula exists*, and we do not have the tedium of listing all the possible answers if all we want to know is how many there are. The number of possible bridge hands of 13 from a deck of 52 cards is $^{52}C_{13} = 635\,013\,559\,600$, and Chapter 2 uses the calculation $^{49}C_6 = 13\,983\,816$ for the number of different ways to fill out a selection on the National Lottery.

The numbers we have calculated are often displayed as shown in Table AI.2, known as Pascal's triangle. Successive rows are labelled 0, 1, 2, 3, ... where the label denotes the number of objects available for selection. Each row, from left to right, lists the number of ways of choosing 0, 1, 2, 3, ... objects in turn.

Table AI.2 Part of Pascal's triangle.

```
                    1
                  1   1
                1   2   1
              1   3   3   1
            1   4   6   4   1
          1   5  10  10   5   1
        1   6  15  20  15   6   1
      1   7  21  35  35  21   7   1
    1   8  28  56  70  56  28   8   1
```

For example, the seventh row 1 6 15 20 15 6 1 applies to choosing from a collection of six objects. Each row is symmetrical about its middle, and for good reason. That row indicates there are 15 ways to select two objects from a collection of six; but whenever two are chosen out of six, that isolates the four not chosen, so there must be exactly the same number of ways to choose four objects from six as to choose two. A similar analysis applies throughout the triangle. Plainly, there is just one way to select six objects from six—you choose all of them; and similarly, there is just one way to choose zero objects—you reject all of them. Every row in the triangle begins and ends with a 1.

Table AI.2 can be extended one row at a time to as many rows as you like without directly using Result 3. Look at any value not at the extremities of a row in the triangle. You will notice that it is equal to what you get when you add the two numbers immediately to the NW and NE in the row above. And this holds no matter how far we extend Pascal's triangle. (In an idle moment, you might work out why this must be so.) So to get the next row, insert the figure 1 at each end, and use this observation to find the intermediate values. In this way, you could work out any value $^{K}C_{r}$, no matter how large K is, given enough time.

Take a different problem. Toss a coin five times, and decide how many sequences, such as HHTHT, could arise. The answer is 32, found by repeated use of the central principle. At each toss, there are just two possibilities, H or T, so when a coin is tossed twice, there are $2 \times 2 = 4$ sequences, HH, HT, TH, TT. A third toss gives two choices to extend each of these, leading to $4 \times 2 = 8$ sequences, and so on. Each extra toss just doubles the number of sequences, so four tosses lead to $8 \times 2 = 16$, and five lead to $16 \times 2 = 32$, as claimed.

Result 4

Tossing a coin n times leads to 2^n distinct outcomes.

The same argument can be applied to a sequence of throws of a die. Since there are six faces, each throw has six outcomes, so two throws have $6 \times 6 = 36$ possibilities. Three throws will then have $6 \times 6 \times 6 = 216$ outcomes, and so on.

Result 5

Suppose an experiment has a fixed number, K, of outcomes. If it is repeated n times, and the results listed in order, there are K^n different sequences.

As we increase n, the number of repetitions, the values in both Result 4 and Result 5 also grow very rapidly. Even for a coin with just two possible outcomes on any toss, a sequence of 20 tosses has over one million distinguishable outcomes. For a die, the million mark is left well behind after just eight throws. *The ability to count the outcomes, without listing them, makes progress possible.*

As an extended example of using these principles, here follows a justification of the figures quoted in Chapter 10 for the numbers of different poker hands.

A poker hand has five cards, which gives $^{52}C_5$ hands in total. Now

$$^{52}C_5 = \frac{52 \times 51 \times 50 \times 49 \times 48}{5 \times 4 \times 3 \times 2 \times 1} = 2\,598\,960.$$

First, count the One Pair hands. The 'pair' can be any one of 13 ranks, A, K, ... , 2; whichever is selected, there are four cards of that rank, so they can be chosen in $^4C_2 = 6$ ways. This gives $13 \times 6 = 78$ ways to choose the pair. The other three cards must be of different ranks, from the 12 remaining ranks, so the ranks can be chosen in $^{12}C_3 = 220$ ways; for each rank, we can pick a card from any of the four suits, so these three cards can be chosen in $220 \times 4 \times 4 \times 4 = 14\,080$ ways. Combining these choices with the 78 ways of selecting the pair gives a total of $78 \times 14\,080 = 1\,098\,240$ One Pair hands.

Now move to Two Pairs. The two ranks for the Pairs can be chosen in $^{13}C_2 = 78$ ways, and there are again $^4C_2 = 6$ ways to select each Pair. That gives $78 \times 6 \times 6 = 2808$ ways to choose those four cards. The last card can be any one of the 44 from the remaining ranks, so the total number of ways is $2808 \times 44 = 123\,552$.

For Three of a Kind, the Three can be one of 13 ranks, and those three cards can be chosen in $^4C_3 = 4$ ways; this gives $13 \times 4 = 52$ ways to select the triple. The other two cards are selected from the other 12 ranks

($^{12}C_2 = 66$), with four choices of suit each time, leading to a total of $52 \times 66 \times 4 \times 4 = 54\,912$ hands.

For a Full House, we can choose the rank of which we hold three cards in 13 ways, and those three cards in four ways. That leaves 12 choices for the rank where we have two cards, and again six ways to choose those two cards, leading to $13 \times 4 \times 12 \times 6 = 3744$ hands.

For Four of a Kind, we just combine all four cards in one of the 13 ranks with any one of the other 48 cards to give $13 \times 48 = 624$ hands.

That deals with all hands in which at least two cards of the same rank are selected. In the other hands, all ranks are different. First, count the Straights. There are ten types of Straight, from A2345 to 10JQKA. For any one of these, there are four choices of suit at each rank, giving $4 \times 4 \times 4 \times 4 \times 4 = 1024$ combinations. But four of these combinations have all four suits the same, so these hands are actually Straight Flushes. This means there are $10 \times 4 = 40$ Straight Flushes, and just $10 \times 1020 = 10\,200$ ordinary Straights.

Each of the four suits contains $^{13}C_5 = 1287$ Flushes; ten of these are also Straights, leaving 1277 ordinary Flushes in each suit, and so $1277 \times 4 = 5108$ Flushes.

All the other hands are worse than One Pair, having just a high card. To find the number of AK hands, there are $^{11}C_3 = 165$ ways to choose the other three ranks. However, the choice of QJ10 gives a Straight, so that leaves just 164 selections. There are five different ranks in these hands; we have seen there are 1020 ways to select the suits while avoiding Flushes, so the total number of AK hands is $164 \times 1020 = 167\,820$.

One way to finish from here is to note that as we know the total number of poker hands altogether, and the number in each category of AK and higher, we can find the number of hands worse than AK by subtraction. A better method is to calculate this number directly, for then we can sum all our answers as a check on our methods. This we now do.

If a hand has an Ace, but no King, the other four ranks can be selected in $^{11}C_4 = 330$ ways; one of these is the choice 2345 that would give a Straight, leaving 329. There are similarly 329 ways to have a King but no Ace (we must avoid choosing QJ109); and there are $^{11}C_5 = 462$ ways of choosing five ranks from Queen and below; the seven selections 23456,..., 8910JQ are Straights, leaving 455 non-Straights. This gives $329 + 329 + 455 = 1113$ ways to choose the ranks; and again there are 1020 ways to select the suits without obtaining a Flush, so the number of hands worse than AK is $1113 \times 1020 = 1\,135\,260$. This is exactly the number needed as an independent check on the other totals. (What a relief!)

Probability

When it is reasonable to assume that all possible outcomes of an experiment are equally likely, then the probability of each outcome is known as soon as we have counted how many there are. This applies to two outcomes from coins, six from dice, 37 from roulette wheels, and so on.

Result 1

When an experiment has N equally likely outcomes, the probability of any one of them is $1/N$.

Often these outcomes are labelled $\{1, 2, 3, \ldots, N\}$, and the phrase *uniform distribution* is in common use for this model. In a uniform distribution, the probabilities of the distinct outcomes are all equal, so we just have to count them.

As well as using the probabilities of single outcomes, we often group outcomes together. For example, in throwing one die, the outcomes $\{2, 4, 6\}$ naturally group as corresponding to 'an even number', the list $\{1, 2, 3, 4\}$ covers 'at most 4'. The word *event* means either a single outcome, or any group of outcomes.

Result 2

In a uniform distribution with N equally likely outcomes, suppose some event corresponds to exactly K of these. The probability of that event is K/N.

Take the case of tossing a fair coin five times. Appendix I showed there are $2^5 = 32$ distinct outcomes such as TTHTT. What is the probability of exactly three Heads from these five tosses?

When we have selected the three places where the Hs will sit, the two Ts inevitably occupy the other two places, so the number of sequences with three Heads is the number of ways of selecting three positions out of five, $^5C_3 = 10$. That gives the probability we seek as 10/32.

As another example, find the chance that your bridge hand contains exactly four Hearts. We calculated in Appendix I how many possible bridge hands there are; now we count how many of them have exactly four Hearts. These Hearts can be selected in $^{13}C_4 = 715$ ways. You also hold nine other cards, drawn from the 39 non-Hearts. They can be chosen in $^{39}C_9 = 211\,915\,132$ ways. Using the central principle of counting, there are $^{13}C_4 \times {}^{39}C_9$ hands that contain four Hearts and nine other cards, and so the probability you hold exactly four Hearts is

$$\frac{^{13}C_4 \times {}^{39}C_9}{^{52}C_{13}} = 0.2386 \text{ to four significant figures.}$$

If you play bridge a lot, you may dream of holding a very long suit; plainly we have the machinery to find the chance you hold not only four Hearts, but even nine Hearts, say. The answer must be

$$\frac{^{13}C_9 \times {}^{39}C_4}{^{52}C_{13}} = 0.0000926$$

or a bit less than one in ten thousand.

Suppose one card is drawn at random from a well-shuffled deck, so that each card has probability 1/52 of being selected. Then

probability of a Spade = 13/52

probability of a Club = 13/52

so probability of a Black card = probability of Spade or Club = 26/52 = 1/2. But note the calculations

probability of Spade = 13/52

probability of King = 4/52

probability of Spade or King = 16/52

and not 17/52 as you might carelessly have written, if you had forgotten that the King of Spades is both a Spade and a King, so should not be counted twice. There are indeed 13 Spades and 4 Kings, but only 16 cards that qualify as 'Either Spade or King'. When looking at composite events 'either A or B', do not forget to check whether A and B have any overlap. When there is no overlap, as in whether a card is a Spade or a Club, the probability of the composite event is found by simple addition of the individual probabilities; the phrase normally used is 'disjoint (or exclusive) events'.

Result 3

If A and B are *disjoint* (*exclusive*) events, the probability of either A or B is the sum of their individual probabilities. If they are not disjoint, the probability will be *less* than this sum. (This result holds however the probabilities are found; it is not restricted to the uniform distribution.)

We saw above that the chance a bridge hand contains four Hearts is about 0.2386. By the same argument, the chance it holds four Spades is also 0.2386. But since the events 'Contains four Hearts' and 'Contains four Spades' are *not* disjoint (there are hands with both four Hearts and four Spades), the chance of 'Either four Hearts or four Spades' is *less than* 0.2386 + 0.2386. But the events 'Contains nine Hearts' and 'Contains nine Spades' *are* disjoint, since there is not room in 13 cards for both of these to happen. The other possibilities of nine Diamonds, or nine Clubs, are disjoint for the same reason, so taking all four mutually disjoint cases together, the chance a bridge hand has *any* nine-card suit is $4 \times 0.0000926 = 0.00037$; about one hand every 2700.

The idea of independence was introduced in Chapter 1, but is repeated here for completeness. If we draw one card from a pack, then

the chance of a Spade is $13/52 = 1/4$

the chance of a King is $4/52 = 1/13$.

Two events A and B are *independent* if the probability they both occur is the product of their individual probabilities. The only way to have both a Spade *and* a King is to have the King of Spades, whose probability is 1/52. But since $1/52 = (1/13) \times (1/4)$, these two events are indeed independent.

Because there is one King in each suit, any knowledge that the card is a Spade is irrelevant to whether it is a King. Similarly, to be told the card is a King gives no clue as to its suit. This is what independence corresponds to. Usually, the context makes it clear whether two events can be taken as independent. Successive tosses of a coin or die are commonly taken to be independent, but the result of drawing a second card from a pack will *not* be independent of drawing the first.

Suppose you roll four ordinary dice. Either at least one of them shows a Six, or none of them do. Which is more likely? On which side would you bet? To explore this, we can be sure that exactly one of the two events

- at least one shows a Six
- none of them show a Six

occurs. Consequently, *the probabilities of the two events must add up to 1.* This trite observation has the following consequence: if we can find the chance of the second event, then we obtain the chance of the first one by simply subtracting the answer from 1.

To solve the problem by finding the probability that none show a Six, use the independence of the four rolls. On one die, since the chance of a Six is 1/6, the chance of a non-Six is 5/6. Since the four rolls are independent, the chance they all show a non-Six is $(5/6)^4 = 625/1296$. Either all are non-Sixes, or at least one is a Six, so the chance at least one is a Six is $1 - 625/1296 = 671/1296$. Since 671 exceeds 625, it is more likely than not that at least one Six will appear. That was the side to bet on.

Finding this probability by listing the various ways we could have one, two, three or four Sixes would be much more complicated. But finding the chance of no Sixes, and subtracting from one, kills the problem at once. Refer to this as Trick 1 in probability.

This Trick is of immense value, and is used frequently. Any event either happens, or it does not. When you add together the probability that it does happen, and the probability it does not, the sum is always equal to 1.

Non-uniform distributions

On many occasions, we know that not all the outcomes are equally likely. If you throw one die, and ask merely whether you get a Six or not, the two outcomes 'Six' and 'Non-Six' may have probabilities 1/6 and 5/6. How would we then work out the chance of, say, at least four Sixes when ten dice are thrown, or the chance that it takes ten or more throws for the first Six to appear? Both these problems have a common framework, as they relate to a sequence of experiments carried out under more or less identical conditions. Our ideal is what is known as the model of *Bernoulli trials*. We require

(1) Each trial has two possible outcomes; according to context, these may be labelled Success or Failure, Head or Tail, etc.

(2) All trials are independent: the outcome of any one of them has no influence at all on the outcome of any other.

(3) The probabilities of the two outcomes remain the same for every trial.

How many Bernoulli trials are needed to get the first Success? It could be 1, 2, 3, ... —any number at all. What are the respective probabilities?

The only way we need exactly five trials is when we begin with four Failures, and then have the first Success; the five outcomes in order are FFFFS. We took the chance of Success at any trial to be constant: call this chance p, so that the chance of Failure is $1 - p$. Because each trial is independent, the sequence FFFFS has probability

$$(1 - p) \times (1 - p) \times (1 - p) \times (1 - p) \times p = (1 - p)^4 p.$$

In the same way, the chance it takes exactly eight trials will be $(1 - p)^7 p$, and so on.

Result 4

In a sequence of Bernoulli trials, the probability that the first Success occurs on trial number n is $(1 - p)^{n-1} p$, when $n = 1, 2, 3, 4, \ldots$, for ever. (This is called a *geometric distribution.*)

In the case when $p = 1/2$, then $1 - p = 1/2$ also. This arises when we are tossing a fair coin, with Heads corresponding to Success. The successive probabilities here are $1/2$, $1/4$, $1/8$, $1/16$, $\ldots$, corresponding to H, TH, TTH, TTTH, etc. This result is used in Chapter 4, in the discussion of the St Petersburg game.

Notice the general pattern of probabilities in Result 4 as n increases. The next one is found by multiplying by $(1 - p)$ each time, so these probabilities steadily decrease. The largest probability is when $n = 1$, then for $n = 2$, $n = 3$, and so on. If you are waiting for a Success to appear in a sequence of Bernoulli trials, it is *always* more likely to appear on the next trial than any other particular trial. Optimism rules, OK?

You might want to find the chance the first Success takes *at least* a dozen trials. One way is to add together all the probabilities for 12, 13, 14, 15, etc., for ever, but there is a much easier way. The only time you will need at least a dozen trials is when the first 11 trials are all Failures. And this has probability $(1 - p)^{11}$, so *that* is the chance you need a dozen or more trials. No addition necessary, just a slightly different way to state the problem.

Result 5

In a sequence of Bernoulli trials, the probability that the first Success takes *at least n* trials is $(1 - p)^{n-1}$.

As well as asking how many trials to wait for a Success, we may want to know the probabilities of the various numbers of Successes during a given number of trials. Perhaps we throw five dice, and seek the chance we get exactly two Sixes. The best approach is through two stages.

First, note which outcomes correspond to the event of interest. In this last example, the list would begin

SSFFF   SFSFF   SFFSF etc.

i.e. all sequences of length five that have exactly two Successes and three Failures.

Second, find the probability of each outcome, and then add all these probabilities up. The probabilities for those listed begin

$$p \times p \times (1 - p) \times (1 - p) \times (1 - p), p \times (1 - p) \times p \times (1 - p) \times (1 - p), \text{etc.}$$

But every one of this list has the same value, $p^2(1 - p)^3$, as each contains the term p twice, and the term $1 - p$ three times, all multiplied together. We know there are $^5C_2 = 10$ such sequences, and so the chance of exactly two Sixes is $10p^2(1 - p)^3$. The same sort of argument works for any problem along these lines, giving Result 6.

Result 6
In a fixed sequence of n Bernoulli trials, the probability of exactly r Successes is

$$^nC_r \times p^r (1 - p)^{n-r}.$$

This also arises often enough to have a name: the *binomial distribution*.

In the binomial distribution, the number of Successes can be anything in the range from zero to n. As you move along this range, the probabilities increase at first, reach some maximum, then fall away to the end. (This includes the possibility, when p is either very close to zero or very close to 1, that the largest probability is at either endpoint.)

Here are some examples of the binomial in action.

(1) What is the chance that we get exactly three Heads when a fair coin is tossed five times?
Solution: The successive tosses fit the pattern of Bernoulli trials, so identify Heads with Success and Tails with Fail. Then $p = 1/2$, and also $1 - p = 1/2$. The formula in Result 6 for the binomial distribution gives the answer as

$$^5C_3 \times \left(\frac{1}{2}\right)^3 \times \left(\frac{1}{2}\right)^2 = 10 \times \left(\frac{1}{2}\right)^5 = \frac{10}{32}.$$

This comes as a relief, since it is the same answer as we found when we solved the same problem by a different method after Result 2!

(2) What is the chance that a backgammon player obtains two doubles in her next four throws?
Solution: In backgammon, a pair of dice are thrown together, and the chance they show the same score is one in six (OK?). So if Success denotes

getting a double, then $p = 1/6$ and $1 - p = 5/6$. The chance of exactly two doubles, using Result 6, is

$$^4C_2 \times \left(\frac{1}{6}\right)^2 \times \left(\frac{5}{6}\right)^2 = 6 \times \frac{1}{36} \times \frac{25}{36} = \frac{150}{1296}$$

(The fraction 150/1296 simplifies to 25/216, about 11–12% but it is seldom useful to take such steps until the end of the problem.)

If the phrase 'two doubles' means exactly two doubles, our task is over, but it might have meant 'at least two doubles'. To complete the tale in this case, the chance of exactly three doubles is

$$^4C_3 \times \left(\frac{1}{6}\right)^3 \times \left(\frac{5}{6}\right)^1 = \frac{20}{1296}$$

and that for four doubles is

$$^4C_4 \times \left(\frac{1}{6}\right)^4 \times \left(\frac{5}{6}\right)^0 = \frac{1}{1296}.$$

Altogether, the chance of two or more doubles is the sum of these three results, as the outcomes are disjoint. The value is

$$\frac{171}{1296} = \frac{19}{144},$$

about 13%.

(3) What is the chance that a double-glazing salesman makes at least two sales in his next ten attempts if, on average, one pitch in ten succeeds?
Solution: We should check on the Bernoulli assumptions. Each attempt ends in either Success or Failure; it does seem reasonable to assume the trials are independent, if he is a keen salesman driven by commission on each sale; and unless his energy flags, his chances of Success are likely to remain constant.

If all that is satisfactory, Result 6 shows that we should use the binomial distribution, and find the chance of exactly r Successes as

$$^{10}C_r \times \left(\frac{1}{10}\right)^r \times \left(\frac{9}{10}\right)^{10-r}.$$

To work out how often he will get at least two Successes, one way is to evaluate this for $r = 2, 3, 4, \ldots, 10$, and then add up the answers. But Trick 1 is to first work out the chance something does *not* happen. Here, either he makes at least two sales, or he makes either zero or one sale. The chances of no Successes, or just one Success are

$$^{10}C_0 \times \left(\frac{1}{10}\right)^0 \times \left(\frac{9}{10}\right)^{10} = \left(\frac{9}{10}\right)^{10} \quad \text{and } ^{10}C_1 \times \left(\frac{1}{10}\right)^1 \times \left(\frac{9}{10}\right)^9 = \left(\frac{9}{10}\right)^9$$

which sum to about 0.7361. So the chance of at least two sales is $1 - 0.7361 = 0.2639$, say 26%. Trick 1 led to two probability calculations, one addition, and one subtraction; the direct method would have needed nine probability calculations and eight additions. Learn to be lazy.

(4) In Chapter 8, we look at how the number of sets played between two players affects their chances of winning a match.
Solution: Suppose Lily's chance of winning any set is $\frac{1}{2} + y$, and sets are played independently. The number of sets she wins in a match of given length will have a binomial distribution. To find her chance of winning a match, we add up all the chances that correspond to her winning a majority of the sets. We find:

Best of three: the chance is $\frac{1}{2} + \frac{3y}{2} - 2y^3$.

Best of five: the chance is $\frac{1}{2} + \frac{15y}{8} - 5y^3 + 6y^5$.

Best of seven: the chance is $\frac{1}{2} + \frac{35y}{16} - \frac{35y^3}{4} + 21y^5 - 20y^7$.

These are exact; only the first two terms matter when y is close to zero.

There is one other named distribution that arises frequently, the *Poisson distribution*, named after the French statistician Simeon Denis Poisson. Suppose we are looking for something that arises quite rarely, but at some average rate: we might be scouring a patch of grass for four-leaf clovers; a garage attendant feeling a surge of excitement when a Porsche fills up with petrol; a second proof-reader marking the misprints not spotted earlier. The most infamous example takes the numbers of Prussian officers kicked to death by horses in various Army Corps over 20 years, during the nineteenth century. The Poisson also arises in place of the binomial when the number of 'trials' is enormous, and the chance of Success at any trial is tiny. For example, in a newspaper 'Spot the Ball' competition, a large but unknown number will enter, each with a small chance of Success. The actual number who guess correctly will tend to follow a Poisson distribution around some average number of winners. If we use X to denote the quantity of interest, suppose its average value is μ. The X will have a Poisson distribution when it can take any of the values 0, 1, 2, 3, ... , and the probability that X takes the value r is

$$\frac{e^{-\mu}\mu^r}{r!}.$$

To see a typical pattern, take the example when $\mu = 2.5$. Then this formula shows the first few probabilities in Table AII.1.

Table AII.1. Probabilities for a Poisson distribution when $\mu = 2.5$.

Value of r	0	1	2	3	4	5	6	At least 7
Probability	0.082	0.205	0.257	0.214	0.134	0.067	0.028	0.014

Like the binomial, the probabilities in a Poisson distribution increase to some maximum, then fall away.

The uniform, the geometric, the binomial and the Poisson arise typically when something is being *counted*. But when we *measure* an exact time, or distance, or any such continuous quantity, they are not appropriate. The probability that Peter weighs *exactly* 72 kg is zero. Even if the scales indicate 72 kg, that is at best a good approximation. With a finer dial, we might decide the weight is really 71.8 kg. Take a magnifying glass, and this gets corrected to 71.82 kg, and so on. To deal with outcomes such as times, weights, and distances, that can vary continuously, we need a different approach.

Rather than speak of the probability that a randomly selected adult male weighs 72 kg, it is more sensible to fix a small interval around that value, perhaps 71.5 to 72.5, or 71.9 to 72.1, depending on how precisely we want to measure. It is perfectly meaningful to consider the probability the weight lies inside such an interval, rather than being *exactly* some value. Here, instead of listing the possible values and giving a probability to each, a graph is drawn and the *area* under the graph corresponds to the probability.

For example, suppose data are collected on ordinary adult males, and a histogram of their weights formed. Each weight will have been recorded as being in some interval such as 71.5 to 72.5 kg, and a smooth curve can be superimposed on the histogram to iron out the effects of sampling fluctuations, as in Figure AII.1. This smooth curve is expected to show the properties of the population from which we have been sampling. The scale is chosen so that the total area under the curve, and above the axis, is one—just as the total of all probabilities must add up to one. Then the probability for any particular interval—say from 70 to 75 kg—is just the area under the curve between those limits, as shown in Figure AII.2.

The most useful of these continuous distributions is the so-called *normal* family, shown by the bell-shaped curve of Figure AII.3. This particular figure shows a graph of the 'standard' normal distribution, and all normal distributions are related to this in a straightforward way. For example, suppose that the weights of adult males in Figure AII.2 follow some normal distribution. Then their graph will be centred around the

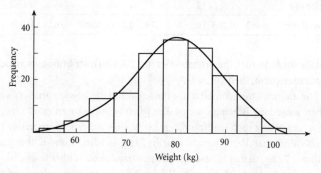

Fig. AII.1 Distribution of weights of 180 adult males (fictional data), with smooth curve superimposed.

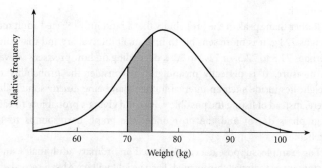

Fig. AII.2 Smooth continuous curve showing what the distribution of the data in Figure AII.1 might settle down to, if sufficient data were collected. The shaded area represents the probability that the weight is between 70 and 75 kg.

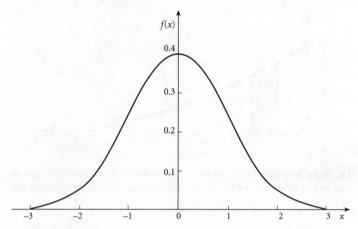

Fig. AII.3 The standard normal distribution.

average weight of these males, and the breadth of the curve will indicate how variable about this average the weights are. By subtracting this average weight, the graph now becomes centred round zero, as in Figure AII.3. And now scaling the breadth according to the units being used, Figures AII.2 and AII.3 become identical.

In this way, using widely available statistical tables for the standard normal distribution, *any* probability for any normal distribution can be found. We have no cause to reproduce these tables; it is enough to quote from them where necessary. The main point is that probabilities for normal distributions are as easy to find as for any of the other named distributions seen earlier. Appendix III contains more information, after we have looked more closely at the notions of average and variability. Whenever continuous data have some well-defined central value, and a fairly symmetrical spread about that value, some normal distribution will usually provide a good fit, and a useful model. It is difficult to overestimate its importance.

We saw that the number of Bernoulli trials until the first Success will follow a geometric distribution. One continuous analogue is waiting for a goal in a soccer match; if goals tend to come along at random, at some average rate, the waiting time will tend to follow a graph similar to that shown in Figure AII.4, the *exponential distribution*. As with calculations for a normal distribution, the scale is such that the total area under the curve is 1, and the probability we have to wait between 12 and 20 minutes for the

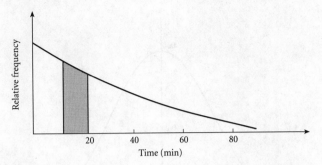

Fig. AII.4 Illustration of the curve for an exponential distribution, as applied to how long we wait for the first goal in a soccer match. The shaded area is the probability that we wait between 12 and 20 minutes.

first goal will be the area shaded. Whenever a problem involves waiting for a continuous period of time for something to happen, this exponential distribution is likely to be called upon. Like the geometric distribution, the probabilities in an exponential distribution fall away; it is more likely that what we are waiting for will happen in the next five minutes than any later specific five-minute period.

Averages and variability

Most people know intuitively what an average is. We have relied on that to use the word freely in a descriptive and informal sense, for example when introducing the Poisson distribution. But to see how useful averages can be, and to appreciate their limitations, it is best to have a more precise approach.

A probability distribution tells you everything: what the possible outcomes are, and their respective frequencies. For some purposes, all this detail is essential, but for others it clutters up your mind and you would prefer less comprehensive information, but enough that captures the important features. These might include:

- the single most likely outcome
- what happens, on average
- the variability around that average.

It is helpful to label what we are interested in as X, the value of some reward. Its probability distribution is the list of all possible rewards we could get, along with their corresponding probabilities. Picking out the most likely is done by inspection; if X denotes the size of your prize when you back a horse in the Grand National, the single most likely value is zero.

Averages

The simplest case worth looking at is when X can take just two possible values, perhaps $X = 2$ or $X = 8$. The *average* reward depends on how likely each value is. Suppose the value 2 arises 70% of the time, so that a typical sequence of rewards could begin

$$2 \ 2 \ 2 \ 8 \ 2 \ 2 \ 8 \ 8 \ 2 \ 2 \ 2 \ 2 \ 8 \ 2 \ 8 \ 2 \ 2 \ 2.$$

The average of these numbers will be much closer to 2 than to 8, because 2 is much more frequent. A sequence of 1000 rewards would have about 700 2s, and 300 8s, so the average would be

$$\frac{700 \times 2 + 300 \times 8}{1000} = 0.7 \times 2 + 0.3 \times 8 = 3.8.$$

This leads to the general way to compute an average for any reward X: *multiply the different values by their respective probabilities, and add up all the terms.* The notation $E(X)$, pronounced 'the expected value of X' is usually used. But to conform to everyday language, I will retain 'average' for this quantity.

Example 1
Find the average score on one throw of a fair die.

Solution

The score, X, takes the values 1, 2, 3 ... , 6, each with probability 1/6. Using the italicized recipe, we find

$$E(X) = 1 \times \frac{1}{6} + 2 \times \frac{1}{6} + 3 \times \frac{1}{6} + 4 \times \frac{1}{6} + 5 \times \frac{1}{6} + 6 \times \frac{1}{6} = 3.5.$$

In both this example and the previous one, the average $E(X)$ is different from any of the possible values that X can take. This is a common occurrence, but some people do fret about it, apparently believing that an average ought to be one of the possible values. This is false: it is quite plausible that families have 2.4 children, on average; and over 99% of Belgians have more legs than average.

Averages are very user-friendly. Double all the possible rewards, and you double their average. Even better, *the average of a sum is the sum of the averages.* That innocent phrase saves the clear thinker many hours for productive use. Compare the two ways to solve the next problem.

Example 2
Find the average total score when two dice are thrown.

First solution

The possible totals for two dice are 2, 3, 4, ... , 11, 12, but they are not equally likely. For instance, labelling the dice as red and blue, the total 4 can arise in three ways as (1, 3), (2, 2), or (3, 1); 5 could come in four ways as (1, 4), (2, 3), (3, 2), or (4, 1); 10 also has three ways, (4, 6), (5, 5), or (6, 4). You could count them all up, work out the respective probabilities, and use the italicized formula. Good luck. If you make no errors, your answer will be 7.

Second solution
Call the score on the red die X, and that on the blue die Y. We want the average value of the total score, $X + Y$. Since the average of a sum is the sum of the averages, and each of X and Y has average 3.5, the average of the total is $3.5 + 3.5 = 7$. *Much* easier.

Example 3
Your possible rewards from one play of a game are 1, 2, 3, with respective probabilities 1/2, 1/3, and 1/6. What is your average reward from 30 games?

Solution
First find the average for one game. Then knowing that the average of a sum is the sum of the averages, realize that the average reward over 30 games is just 30 times the average for each game.

On one game, if the reward is X,

$$E(X) = 1 \times \frac{1}{2} + 2 \times \frac{1}{3} + 3 \times \frac{1}{6} = \frac{5}{3}.$$

So, over 30 games, the average reward is $30 \times \frac{5}{3} = 50$.

The binomial distribution was introduced in Appendix II, Result 6. Its average value can be found directly, without using the formula given there. This distribution arises when we ask how many Successes there are in n repetitions of an experiment, where the chance of Success is fixed at p each time.

Take the easy case: $n = 1$. If we make just one trial, either we have one Success with probability p, or none with probability $1 - p$. So the average works out as

$$1 \times p + 0 \times (1 - p) = p.$$

Repeating this experiment n times, with an average of p Successes each time, gives an overall average of np. So that must be the average for our binomial.

Example 4
Let's check this out with a specific example. Take four trials, with the chance of Success fixed at 1/3 each trial. According to this slick argument, the average number of Successes will be $4 \times (1/3) = 4/3$.

Using the formula for this binomial, and suppressing the gory details, the values 0, 1, 2, 3, and 4 have probabilities 16/81, 32/81, 24/81, 8/81, and 1/81, so the average will be obtained by multiplying the values by the corresponding probabilities, and adding them all up. (Just thinking of doing it this way brings on a migraine attack.) The final answer is 108/81 = 4/3.

I know which method I prefer.

Examples 5, 6

Chapter 1 invited you to estimate how many of the eight fixtures you might guess correctly, if 16 soccer teams are paired at random. Chapter 7 introduced a card-matching problem, asking you to guess how many pairs of cards will match exactly when the order of two well-shuffled packs is compared. Both these problems can be solved by this *sum of the averages is the average of the sums* idea.

For the cards, look at any specific position—say the 23rd. *Ignore the rest of the cards.* Give yourself the score of 1 if the two cards match, and 0 if they do not. Do the same for every position, and add up all the scores. The total is obviously the number of times you get a match. At any position, you get a match with probability 1/52, so your average score *at that position* is

$$1 \times (1/52) + 0 \times (51/52) = 1/52.$$

It is the same at all 52 positions, so the sum of these averages is $52 \times (1/52)$ = 1. That means the average of the sum is 1. In other words, the average number of matches is 1. Plainly, the same argument works no matter how big the packs are.

For the soccer games, fix attention on the eight pairings that arise in the real draw, and in each one, note the name of the team that is first in an alphabetical listing. (This is just to make sure there is no double counting.) Give a score of 1 if we have guessed the correct opponent for this team, otherwise score 0. Adding up the scores over the eight games just counts how many pairings we guessed correctly. There are 15 possible opponents, so the chance of a correct guess at any game is 1/15. By the same argument as with the cards, the *average* score for each game is 1/15. As there are eight games, the total average is 8/15. And that is why, in Chapter 1, for a reward of £1.50 per correct guess, an entrance fee of 80p is fair.

This 'average of a sum is the sum of the averages' makes mincemeat of some problems that look gruesome if you had to work with the definition of an average.

Follow-up

The answer, 8/15, comes from the facts that there are eight pairings, and any team has 15 possible opponents to guess from. So had we asked the question for a random draw between 24 teams, the average number of correct answers would be 12/23. At the semi-final stage, with just four teams left, the average is 2/3. This last result is easy enough to work out directly: one time in three we correctly guess both pairings, the rest of the time we get both wrong. It is always a comfort to have a check on an answer.

Result 4 in Appendix II describes the geometric distribution, which applies when your concern is how many trials to record the first Success. Its average is of direct interest—if your chance of winning any game is 10%, say, how many games do you play, on average, to record the first win?

Example 7

For the geometric distribution, the average number of trials to the first Success is $1/p$.

Justification

Write $q = 1 - p$ for convenience. We need 1, 2, 3, 4, ... trials with respective probabilities p, qp, q^2p, q^3p, ... and so the average, using the definition, is

$$E(X) = 1 \times p + 2 \times qp + 3 \times q^2p + 4 \times q^3p + \ldots$$
$$= p(1 + 2q + 3q^2 + 4q^3 + \ldots)$$

A piece of A level maths shows that when q lies between zero and 1, the sum in brackets collapses down to $1/(1 - q)^2$. Given this, since $1 - q$ is just p, the average is $1/p$, as claimed.

With a fair coin, then $p = 1/2$, so the average number of tosses to get the first Head is two. And if the chance of Success at any trial is 1/4, it takes, on average, four trials to get a Success. These results are in accord with most people's intuitive idea about probability. For the Lottery jackpot, since your chance on any ticket is $1/13\,983\,816$, you need to buy $13\,983\,816$ tickets, on average, to win a share for the first time. Succinctly, if things happen at rate p, the average time you wait is $1/p$.

The definition of the Poisson distribution in Appendix II includes the statement that its average value is μ. This is another place where A level maths would be needed if we used the formula to work out $E(X)$.

This quantity, $E(X)$, corresponds to the generally accepted meaning of the word 'average', and it is what I mean when I use that word. There is an associated, but different, quantity, the *median*, which is sometimes more useful. The median in a collection of rewards is the middle value, when they are put in order, and taking account of the frequencies. Half the time the reward is no more than the median, half the time it is at least that value. If this 'reward' denotes the salaries in the organization you work for, some very large salaries paid to a small number of employees may inflate the value of $E(X)$, without making a difference to the median.

Take the example of seven equally likely values $\{1\ 2\ 3\ 4\ 5\ 6\ 49\}$. The median is 4: a random selection is equally likely to be above it as below it, and the median remains at 4 whether the highest value is 49, 7, or seven million. But this highest value alters the average greatly; $E(X)$ is just the sum of the values, divided by 7. As the numbers stand, the average is 10, altering the highest value to 7 makes the average 4; altering it to seven million makes the average exceed one million.

Sometimes the list of rewards is completely symmetrical about some middle value. In this case, the median and $E(X)$ are the same. And if there is near-symmetry, the two values will be close.

Variability

The fact that the average payout on one ticket in the National Lottery is 45p is simultaneously

- the central dominating figure for the Lottery
- almost irrelevant

depending on your perspective. To the organizers, the average payout is crucial. To the typical small gambler, losing an average of 55p each week is not noticed. Changing the prize structure so that all entrants are solemnly awarded 45p leaves this average outcome the same, but produces a Lottery that no rational person would enter. It is the variability in the outcomes that offers a game at all.

There are several ways of calculating a single number that might fairly be thought to describe variability. The most common leads to the *variance*. In a phrase, the variance is 'the average squared deviation from $E(X)$', and it has a mathematical formula found in any statistics textbook. Staying with the paradigm that our item of interest, X, is some reward with an average value $E(X)$, the variance will be *small* when it is very likely that the actual reward will be quite close to the average, and it will be *large* when some possible rewards are a good distance from the average.

The smallest possible value for a variance is zero. And that can only arise when there is no variability whatsoever, it is 100% certain that the reward takes a definite single value. Otherwise, a variance is positive. Just as probabilities always sit between zero and 1, so negative variances cannot occur.

We can also find the variance of each of the distributions we have given a name to. Using the same notation as earlier, the variance for the binomial distribution is $np(1 - p)$; that of the geometric is $(1 - p)/p^2$; and for the Poisson, μ is also its variance.

As the first step in computing a variance is to *square* the differences from the average, a variance is not measured in the same units as the original reward. To remedy this, taking the square root of a variance leads to the *standard deviation* which is now measured in the original units. Because variance and standard deviation are square and square root of each other, they carry exactly the same information. A variance is easier to compute, a standard deviation is easier to interpret. Either of them is often quoted as 'the' measure of variability, but they cannot do justice to all the ways variability can be measured. They are most useful when

(a) the rewards are fairly symmetrically distributed about the average

(b) the chances of the different rewards get smaller the further they are from the average.

The binomial, Poisson and normal distributions from Appendix II tend to have these properties, but the geometric and exponential, on their own, do not.

As noted in Appendix II, all probabilities related to a normal distribution can be found when we know its average and its standard deviation. Three benchmarks from normal statistical tables are useful as reference points.

(1) The chance of being within *one* standard deviation of the average is about two-thirds.

(2) The chance of being within *two* standard deviations is about 95%; only once in 20 times is the value further away.

(3) The chance of being within *three* standard deviations is about 99.7%; only once in 370 times are we further away.

See Figure AII.3.

Consider the total reward from a series of games. We have seen that the average of any sum is the sum of the averages. That led to the knowledge that the average total reward is just the sum of the individual averages. One reason why variances are user-friendly is the parallel result: *the variance of the sum of (independent) rewards is the sum of the individual variances.*

There is one final piece for this jigsaw: whenever a total reward is the sum of many independent quantities, it is often the case that (a) and (b) above hold for the whole sum, *even if they are false for every component of the sum.* And when (a) and (b) hold, the benchmarks quoted for the normal distribution give a good approximation.

This is the reason why, in Chapter 7, we felt able to use the benchmarks for a normal distribution when estimating how many cornflakes packets we need buy to obtain a complete set of enclosed coupons. The justification for using them is that the total number of packets needed for the whole collection is the sum of the numbers needed to acquire the first, second, third, … , 24th coupons. This is like adding together 24 'rewards'. Neither (a) nor (b) holds for any one of the 24 components, but both conditions hold fairly well for this sum.

In the computation of the variance of the total number of cornflakes packets, it was vital that each component of that sum was independent of the rest. It is plain that the components *are* independent: the information that it took (say) five purchases to obtain the tenth coupon makes no difference to how many extra packets we must now buy to get the 11th— all that matters is that we currently have ten different coupons. For each coupon in the sequence, the actual number of purchases will follow some geometric distribution.

In Chapter 7, we looked at the specific example of asking how many extra purchases we need, once we have the 17th coupon, to obtain the 18th; on average, this number was 24/7. More completely, the number has a geometric distribution with that mean, i.e. a geometric distribution with $p = 7/24$. Hence its variance will be (see above) $(1 - p)/p^2$, or $(17/24)/(7/24)^2$. This simplifies down to $(24 \times 17)/7^2$. A similar calculation applies to all the other components in the sum.

In the general case, where we are seeking to collect a total of N coupons, we saw that the average number of packets we would need to buy was

$$N\left(1 + \frac{1}{2} + \frac{1}{3} + \ldots + \frac{1}{N}\right).$$

Using the independence of the components of this total sum, the variance of the number we should buy is (exactly)

$$N^2\left(1 + \frac{1}{2^2} + \frac{1}{3^2} + \ldots + \frac{1}{N^2}\right) - N\left(1 + \frac{1}{2} + \frac{1}{3} + \ldots + \frac{1}{N}\right).$$

When N is at least 10, there is an excellent approximation: this variance is close to

$$N^2\pi^2/6 - N(0.577 + 1n(N)).$$

When N is 20 or more, the second term is *much* smaller than the first, and can effectively be ignored. Then, taking the square root leads to our estimate $N \times \pi/\sqrt{6}$, or about $1.28N$, as the standard deviation of the total number of packets to be bought. For example, with $N = 49$, as in the bonus numbers for the National Lottery (Chapter 7), the mean time for all of them to appear was calculated as about 219 draws. We use the above to estimate the standard deviation as a little less than $1.28 \times 49 = 63$, and so the actual time to wait, 262 draws, is under one standard deviation from its average.

Similarly, we can assess the likely range of outcomes from a bout of roulette playing (Chapter 10). Two hundred spins may be reasonably representative of an afternoon session at the wheel, and should be large enough to help iron out some of the random fluctuations, so that (a) and (b) hold for your total reward. For all the bets, except those on Red, the average outcome from a sequence of 200 unit bets is a loss of $200 \times (1/37)$ = 5.4 units. The variance will depend on which wagers were made.

If all your bets are on columns of a dozen numbers, the variance per spin is 2.70, so the variance over the session is $200 \times 2.70 = 540$, leading to a standard deviation of 23.3, its square root. One standard deviation either side of the average ranges from a loss of about 29 units to a profit of 18. This happens in about two-thirds of all sessions, and the rest of the time splits fairly evenly between better and worse. One-sixth of the time, you will finish more than 18 units ahead, another one-sixth you will be more than 29 units down. To make *exact* calculations of the range of outcomes over 200 spins would be quite tedious, but these pointers are reached almost instantly.

If all bets are on a single number, the variance associated with one spin is much larger, 34.08, and the standard deviation over the session computes, in the same way, to 82.6. The size of this standard deviation swamps the average of 5.4; one standard deviation either side of the average runs from a loss of 88 units to a profit of 77 units. One time in three, the actual result will be outside this range. The outcome of this gambling session is very unpredictable, and the average is atypical.

Bets on Red give an average loss of only 2.7 units, and the standard deviation over 200 spins is down to 14.1. Looking at the second benchmark of two standard deviations, in 95% of such sessions your fortune is between a loss of 31 units and a profit of 25. The rest of the time, one time in 20, you fall outside these bounds. The final benchmark of three standard deviations shows how desperately unlucky you would be to be more than 45 units down at the end: that happens only once every 740 ($= 2 \times 370$) visits.

For a gambler over many sessions, the total number of spins will be much larger. The central place of square roots now becomes clear: over N plays, the average loss is $N/37$ units, and if each bet has a variance of V, the total variance is NV. But the standard deviation, its square root, is then a multiple of $\sqrt{N}$; and $\sqrt{N}$ increases much more slowly than N itself. As a *proportion* of the average loss over a period, the standard deviation decreases, and your actual outcome is more predictable.

Take the example of 20 000 spins, where the average loss is seen to be 540 units. For bets on columns the standard deviation has increased to 230. So the second benchmark of two standard deviations either side of the average corresponds to the loss range of 80 to 1000. This will happen 95% of the time. If you play for 100 sessions with 200 spins in each, you will have won in quite a few, but your overall chance of being ahead is only about 1%.

But on the Lottery, Pools or Premium Bonds, the reward variance is so large that these benchmarks will apply only to an *enormous* number of plays, more than one person will be exposed to in a lifetime. Here the benchmarks for the normal distribution cannot be used with any confidence, and you have to rely on exact calculation.

There is no one-line answer to the plaintive query 'When can we use these normal guidelines when adding up a lot of rewards?'. The more symmetric and the less extreme the distribution of rewards on one play, the quicker they become adequate. Contrast betting on Red at roulette with buying a Lottery ticket. There is more to 'variability' than a single formula.

Goodness-of-fit tests

How should we decide whether a particular coin is equally likely to show Heads or Tails? Or whether the dice in your Monopoly set are fair? The logic behind making decisions on these and similar questions is identical, although the details differ. The principle is that a device is presumed innocent until proved guilty by the data it generates. We compare these data with what we expect from a fair device, and so long as they are compatible with fairness, we have no evidence of any bias. Take it that you know in advance how much data will be available. The five steps in the statistical procedure are then:

(1) assume your device (coin, die, Lottery machine, etc.) really is fair;

(2) work out what should happen, on average;

(3) measure how *extreme* the actual data are, in comparison with this average;

(4) calculate p, the probability that a fair device would give data at least this extreme;

(5) if p is very small, reject the presumption of fairness. Otherwise, agree there is not enough evidence to say the device is biased.

The *smaller* this probability p, the stronger the evidence of bias. A large value of p simply means that the data are *consistent* with the device being fair, which is quite different from a *proof* that the device is fair. Even if a device is biased, the amount of data we have may be insufficient to identify the bias.

Formally, this procedure is termed a *goodness-of-fit test*. To see the steps in action, consider the case where there are just two alternatives, that of tossing a coin. Suppose our data arise from 100 tosses, and we record X, the number of Heads and Y, the number of Tails. Plainly $Y = 100 - X$, but using X and Y preserves symmetry.

Step (1) is to assume the coin is fair, which means that both H and T have probability one-half. Hence the average numbers of H and T are both

50 (step (2)). We need some overall measure of extremity. A first instinct is to look at the differences between the actual values X, Y and their averages, but two thoughts intrude:

(1) Whenever there are too few H, there are too many T, so if we just add up the differences, they will cancel out as one is negative and the other positive. To avoid this, we can *square* the differences before adding them up.

(2) Compared with an average of 50, a value of, say, 80 seems a long way away. On the other hand, if the average is 1050, a value of 1080 looks quite close, although the differences are the same. Our measure of extremity needs to take account of this, and one way is to divide these squares by the average. This does the right sort of thing, as it scales a given difference down when the average is already big.

This leads to the notion that a single number that measures the extremity of the data is

$$W = \frac{(X - 50)^2}{50} + \frac{(Y - 50)^2}{50}.$$

Computing W, the *goodness-of-fit statistic*, is step (3). When both X and Y are equal to their average, then W is zero, and the further they are from average, the bigger W gets. Small values of W correspond to a fair device.

For step (4), we need to know what values of W to expect from a fair coin. If such a coin is tossed 100 times, the number of Heads follows a binomial distribution (Appendix II), so we can calculate the possible values of W, and their respective chances.

For example, suppose there are 55 Heads. Then $X = 55$, $Y = 45$, and W evaluates to be exactly 1. The probability of this or any larger value turns out to be about 37%. This is quite an appreciable probability, so we should not be alarmed if we get 55 Heads when 50 are expected—a fair coin leads to such an outcome more than a third of the time. $W = 1$ is quite consistent with a fair coin.

But suppose Heads occur 60 times. Then $X = 60$, $Y = 40$ and $W = 4$. The chance of W being 4 or larger is only about 5.7%, so if the coin really is fair, an outcome as extreme as this arises in only one experiment in 18, or so. There are some suspicions about fairness. Press on a little further and imagine there are 65 Heads. Here W is now 9, and the chance of such a large value is down to 1/3%. With a fair coin, only one experiment in 300 gives such an extreme value. This would be good evidence that the coin is biased.

These three calculations for step (4) have led to p being 37%, 5.7%, and 1/3%. There can be no absolute rule that gives a universal cut-off point and declares a coin fair or biased according to which side of the divide p falls. We can only have greater or lesser suspicion. But years of statistical experience over many applications have thrown up some guidelines, which can be summarized as:

(1) If p is above 5%, be reluctant to claim the device is unfair.

(2) If p is between 1% and 5%, agree there is some evidence against fairness.

(3) If p is between 1/10% and 1%, claim there is good evidence against fairness.

(4) If p is less than 1/10%, assert that there is strong (but not conclusive) evidence against fairness.

I emphasize: these are merely guidelines, and not instructions. For a coin with any (large) number of tosses, the values of W that correspond to 5%, 1%, and 1/10% are roughly 4, 9, and 14. Exact values are in Table AIV.1.

Move on to something more complex than a two-sided coin. How could we use 120 tosses of a six-sided die to judge its fairness? Taking the same pattern, step (1) is to note that a fair die has probability 1/6 for each face. Thus, in 120 tosses, each face appears 20 times on average (step (2)). Writing A, B, C, D, E, and F as the numbers of times the six faces appear, the corresponding measure of extremity is

$$W = \frac{(A - 20)^2}{20} + \frac{(B - 20)^2}{20} + \ldots + \frac{(F - 20)^2}{20}.$$

The only time that W is zero is when every face appears exactly 20 times, and the further the frequencies are from average, the bigger W gets. Calculating the probabilities for its different values is more complicated than with the coin, but there are statistical tables, the *chi-square tables*, that enable us to read off these probabilities. With the coin, W arose from adding two quantities together. Here, six are added, and W will tend to be bigger, and have a different distribution. A fair die will give an average value of 5, and anything up to about 11 is not unexpected. The values of W that correspond to the trigger probabilities of 5%, 1%, and 1/10% are 11, 15, and 21. Once again, the bigger the value of W, the more suspicious we are about the die. Strictly speaking, this procedure should be applied only when we have a lot of data, but in practice, so long as all the average frequencies in step (2) are at least 5, these calculations will not mislead.

At the beginning of Chapter 5, there is a small table showing the outcomes of four separate experiments where a biased die was thrown 120 times. Glancing at that data gives no suggestion of a bias, and proper calculations back that impression up. The four values of W are easily found to be 6.8, 4.1, 9.8, and 4.6. Three of them are quite close to the average expected, 5, and none of them exceed the lowest critical value, 11. A mere 120 throws have been too few to detect the bias we know is present.

So far, we have quoted figures for the critical values of W for two different calculations, a coin and a die. The critical values will depend on the average value of W expected for a fair device; this average is called the *degrees of freedom* and is usually (but not, alas, always) one fewer than the number of terms we add together to find W. Rather than give an all-embracing rule to calculate the degrees of freedom that would apply to any use of these tables, I will simply state the answers for the problems we look at.

Finding statistical evidence of a bias can be quite hard. The size of the bias we hope to detect is crucial. Return to a coin, and consider the sort of experiment needed to produce data that will give a good chance of picking up the bias, if the true chance of Heads is 47%. What hope do we have with an experiment of 100 tosses?

The brief answer is: very little hope. If a fair coin is tossed 100 times, the number of Heads falls outside the range 40 to 60 about 5% of the time. Our statistical test will only begin to claim unfairness if the number of Heads is below 40, or above 60. But with this particular biased coin, 100 tosses will average 47 Heads, and 93% of the time the number of Heads will be inside the range that is compatible with fairness. One hundred tosses are nowhere near enough.

Try a more extensive experiment, with 1000 tosses. In the formula that gives W, all the entries '50' change to '500', and now a fair coin shows fewer than 468, or more than 532, Heads about 5% of the time. Any number of Heads between 468 and 532 is quite compatible with fairness. Our same biased coin obviously has an average of 470 Heads. It will fall inside or outside the interval from 468 to 532 about equally often, and so even this experiment has only a 50:50 chance of detecting the bias. But with 10 000 tosses, at last the experiment is sensitive enough. The 'acceptable' range is from 4900 to 5100, and the chance that a coin whose true probability of Heads is 47% will fail to be detected is about one in 30 000.

Be clear what this procedure means. We wish to reach a decision about whether a device is fair, and we reach our conclusion on the basis of some value of W that corresponds to some probability, p. The bigger W, and the smaller p, the more convinced we are of unfairness. But taking $p = 5\%$ means that out of every 100 genuinely fair devices we test, we shall (wrongly) declare five of them biased, as random chance will give an extreme value of W. We never know which five!

Making such decisions, on coins, dice, or whatever, leads to two distinct kinds of mistakes. One is to wrongly reject a perfectly good device, the other is to declare a faulty device acceptable. And, for a given amount of data, if you reduce your chance of making one of these types of error, you inevitably increase your chance of making the other. If you reduce p from 5% to 1%, you will only make a mistake with one fair device in 100, but you will let through far more faulty devices that pass this more lenient test. To make fewer errors of both types you require more data.

For the numbers in the National Lottery, the procedure is similar. In any draw, six different main numbers are selected, so in a fair draw every number has chance 6/49 of being chosen (step (1)). The average number of times a given number is selected in D draws is just $6D/49$ (step (2)). Call this average A. For each of the 49 numbers, count up how often it has appeared, X, say. Then the contribution to W from that number is

$$\frac{(X - A)^2}{A};$$

add all 49 contributions together. But because six numbers are chosen in each draw, one additional step is now needed to use the standard chi-square tables—this sum must be multiplied by 48/43. That completes step (3).

Having found W, we need to know the critical values that would indicate some bias in the Lottery draw. This time, with a fair Lottery, the average value of W would be 48. The critical values for 5%, 1%, and 1/10% are about 65, 74, and 84. When using this procedure, it is recommended that the average, A, should be at least 5; this is achieved when we look at 41 or more draws, since $6 \times 41/49 = 5.02$. Dividing the Lottery draws into non-overlapping blocks of 41, the values of W for the first six blocks, to draw 246, were 35.8, 38.0, 40.2, 59.8, 44.2 and 42.0. All of these are less than that lowest critical value, 65, and just the sort of values you might expect of a fair Lottery.

In deciding whether a coin, die or Lottery machine is fair, it is not enough just to check whether all outcomes are equally frequent. It is also essential to know that successive outcomes are independent of each other. A coin that alternated regularly between H and T would certainly pass any test of equality of frequency, but it is not a random sequence.

To make a statistical test of independence of consecutive outcomes, we could count how often H was followed by either H or T, and also how often T was followed by H or T. Take the example of the sequence of 30 tosses described when we introduce Penney-ante in Chapter 4, repeated here for convenience:

HTHTT THTTT HHTTH HTHTT HTHTH HHHTT.

They show 14 Hs and 16 Ts, so the formal goodness-of-fit statistic for equal frequency is

$$W_1 = \frac{(14 - 15)^2}{15} + \frac{(16 - 15)^2}{15} = \frac{2}{15} = 0.1333$$

which is obviously acceptable. Look just at the 14 Hs: the *next* toss is H five times, and T nine times; and after the 16 Ts we have nine Hs and seven Ts (imagine the final T is followed by a H at toss 31). Write this as a table:

		Next toss:	
		Head	Tail
Present toss:	Head	5	9
	Tail	9	7

If H and T are equally frequent and successive tosses are independent, all four entries in the table will have the same average (steps (1) and (2)); this average is $30/4 = 7.5 = A$, say. That leads to the goodness-of-fit statistic

$$W_2 = \frac{(5 - A)^2}{A} + \frac{(9 - A)^2}{A} + \frac{(9 - A)^2}{A} + \frac{(7 - A)^2}{A} = 1.4667.$$

To take account of any random differences in the overall frequencies of H and T, the goodness-of-fit statistic for independence is the *difference* between W_2 and W_1, so here $W = 1.4667 - 0.1333 = 1.33$ (step (3)).

To come to a decision, we turn to the chi-square tables to see what values of W would be so large that would make us believe lack of independence. In this instance, the average value expected of W turns out to be 2, and the critical values that correspond to 5%, 1%, and 1/10% are (about) 6, 9, and 14. Our value of 1.33 shows no evidence that one toss influences the next.

But if our 30 tosses had strictly alternated between H and T, the four entries in the table would have been 0,15,15 and 0. This time, the value of

W_1 would have been zero, but W_2 would be 30, and so W is also 30. This is quite enormous, in context, and overwhelming *evidence* of lack of independence.

For a similar question with a six-sided die, a parallel path is followed. First, count the frequency of each face, and calculate the goodness-of-fit statistic W_1 that tests equality of frequency among the six faces as normal. Then trawl through the data again, counting how often each face is followed immediately by each of the six possibilities, and construct a 6×6 table. That leads to a calculation of W_2 as a sum of 36 quantities. Finally, the goodness-of-fit statistic W is the difference between W_2 and W_1. Purely for reference, the average value if all tosses are independent is 30, and the chi-square tables show the three critical values of W as 44, 51 and 60. Since there are 36 entries in the table, to have sufficient data to make every average at least 5, we should make at least 180 tosses for this test of independence.

In all these examples, the problem has been to judge whether a number of alternatives is equally likely. When some alternatives are thought more likely than others, the method is very similar. One famous example is the data that Gregor Mendel collected on peas. (The statistical work that would have enabled him to carry out a proper goodness-of-fit test was not done until many years later.) His theory of heredity led him to believe that four different types of pea should be produced in the ratio 9:3:3:1, which give probabilities of

$$\frac{9}{16}, \frac{3}{16}, \frac{3}{16}, \quad \text{and} \quad \frac{1}{16}$$

(step (1)). His data were the actual numbers of the four types. To keep the arithmetic easy, suppose he had 320 peas; then the *average* numbers of the four types would be 180, 60, 60, and 20 (step (2)). To find the goodness-of-fit statistic, for each type calculate the difference between actual and average, square it, divide by the average. Then add these four quantities together to obtain W (step (3)). If Mendel's theory gave the right frequencies, the average value of W would be 3, and the guideline critical values are about 8, 11, and 16. Mendel's actual data fitted his theory beautifully, with W very close to zero (some scientists have claimed the data are *suspiciously good*: rather like tossing a coin 10 000 times and getting 5001 heads). The rediscovery of his work 40 years later laid the foundations of modern biology.

To use these goodness-of-fit tests, we need a table that gives the critical values of W for 5%, 1%, and 1/10% probabilities that correspond to different averages. A brief version can be found in Table AIV.1. (The values quoted in the text were rounded versions of these more exact values.)

Table AIV.1 Some critical values for the goodness-of-fit test.

Average	5%	1%	1/10%	Average	5%	1%	1/10%
1	3.84	6.64	10.63	10	18.31	23.21	29.59
2	5.99	9.21	13.82	15	25.00	30.58	37.70
3	7.82	11.34	16.27	20	31.41	37.57	45.31
4	9.49	13.28	18.47	30	43.77	50.89	59.70
5	11.07	15.09	20.52	40	55.76	63.69	73.40
8	15.51	20.09	26.12	50	67.50	76.15	86.66

The Kelly strategy

In Chapter 11, where we imagine being in a favourable game, a description is given of how to use the strategy described by John L. Kelly Jr to increase your fortune at the maximum safe rate. Here, we give a more mathematical account, showing where this strategy comes from.

A bet at even payout

Your initial fortune is denoted by F, and you are in a game that favours you in the following way. You can bet any amount you actually have, and you either lose that sum, or win an equal amount. But your chance of winning at any bet is p, and p *exceeds* 50%. Suppose your strategy is to bet a fixed fraction, x, of your current fortune at each play. What fraction should you choose?

A greedy player will see a favourable game, and bet large amounts. He wins more often than he loses, but when he loses, his fortune drops dramatically. A really greedy player will stake his entire fortune each time, and will happily double up until one loss wipes him out completely. The timid player bets the minimum each time, increases his fortune, but oh so slowly! What is the happy medium?

You bet a fraction x of your fortune. When you win, you have $1 + x$ times as much as before, when you lose you have $1 - x$ times as much. After N plays, when you have won M times and lost $N - M$ times, your fortune is G, where

$$G = (1 + x)^M (1 - x)^{N-M}.F.$$

If you have some long-term average growth rate, your fortune should increase like compound interest, multiplicatively, and so we can best find this rate by taking logarithms. Divide both sides by F first, to get

$$\log \left(\frac{G}{F}\right) = M \log (1 + x) + (N - M) \log (1 - x).$$

Divide both sides by N. The left side now represents the average growth rate at each play, which is what we want to maximize. But on the right side, the fraction M/N is the proportion of the first N bets that are winning ones. When N is large, this will be close to p. Similarly, $(N - M)/N$ will be close to $1 - p$. The average growth rate, after a long time, is given by

$$p \log (1 + x) + (1 - p) \log (1 - x). \qquad (*)$$

To anyone familiar with calculus, finding the value of x that makes this a maximum is a standard problem, and leads to the solution that $x = p - (1 - p)$. But since p is the chance you win, so $1 - p$ is the chance you lose, and this difference is the size of your advantage. And that is Kelly's first result: *in a favourable game with even payouts, the fraction of your fortune you should bet is just the size of your advantage.*

That strategy maximizes the average rate at which your fortune grows. Bet more, or less, and your fortune will not grow so fast, on average.

For example, when $p = 52\% = 0.52$, then $1 - p = 0.48$, and the value of x is their difference, 4%. You maximize your long-term average growth rate by betting 4% of your current fortune each time. To find what this optimal growth rate is, we work out what the value of $(*)$ is when we choose x optimally; to a very good approximation, especially when p is only just above 50%, the answer is $x^2/2$. Betting this 4% of your fortune leads to a growth rate of 0.0008 on each play, on average. After N plays, you have about 1.0008^N times as much as you started with. But this average will conceal some large fluctuations, as you will have several wins, or losses, in succession while you play.

Bets at other payouts

Suppose the payout is at odds of K:1 (K may be less than, or greater than, 1), and your winning chance is again p. The game is favourable so long as $p \times (K + 1)$ exceeds 1, so we suppose this happens. Using the same strategy as above, the different payout means that if you win, your fortune is now $(1 + Kx)$ times as much as before, while if you lose, it is again $(1 - x)$ times as much. That small change carries through to give the average growth rate that corresponds to $(*)$ as

$$p \log (1 + Kx) + (1 - p) \log (1 - x).$$

A similar piece of calculus means that the best choice of x is $p - (1 - p)/K$. And in this case, the long-term average growth rate is about

$$K \frac{x^2}{2} (1 - x(K - 1)).$$

Thus, for example, if the payout is 2:1 but you have a 40% chance of winning, you should bet

$$0.4 - (1 - 0.4)/2 = 0.1 = 10\%$$

of your fortune each time. Your fortune grows, on average, at 0.008 each bet.

A horse race with several horses

In this race, each horse has some probability of winning, call it p, and it will be offered at some odds, say α to 1. Betting on that particular horse is favourable whenever the product, $p \times (\alpha + 1)$, strictly exceeds 1. In many races, especially where your judgement coincides with the bookie's judgement, that condition will not be satisfied for any horse, and there will be no favourable bet. But sometimes, the race will have one or more horses for which the odds offered are long enough for you to have this advantage. Suppose you are in this happy position: on which horses should you bet, and how much on each?

I will describe how to act in a general race, and illustrate what these steps are for the example described in Chapter 11. The justification for this recipe can be found in Kelly's work: here I aim to describe it fully enough so that a numerate reader can apply it to any race. In our example, the horses are A, B, C, D, and E, with respective probabilities 40%, 30%, 20%, 8%, and 2%, and odds of 13:8, 9:4, 9:2, 5:1, and 20:1.

The first step is to evaluate $p \times (\alpha + 1)$ for every horse in the race, and label the n horses as $1, 2, \ldots, n$ in *decreasing* order of these values. For the first horse, and perhaps the first few horses, this quantity exceeds 1, but near the end it may be quite small. In our example, the five values are $0.4 \times (21/8) = 1.05$, $0.3 \times (13/4) = 0.975$, $0.2 \times (11/2) = 1.1$, $0.08 \times (6/1) = 0.48$ and $0.02 \times (21/1) = 0.42$. So we should order them as C, A, B, D, and E.

Next, fix some value k, and work out two quantities for the first k horses: the first is $P(k)$, the sum of the probabilities that these horses will win, and the second is $A(k)$, the sum of the values of $1/(\alpha + 1)$ for these same horses. When $k = 1$, we just have horse C for which the values are 20% and 2/11; when $k = 2$, we take both C and A and obtain 20% + 40% = 60%, and 2/11 + 8/21 = 0.56277..; when $k = 3$, we have C, A, and B and the sums are 20% + 40% + 30% = 90% and 2/11 + 8/21 + 4/13 = 0.87046..; when $k = 4$, we include D, so the sum of the probabilities is now 98%, and the sum of the reciprocals has another 1/6 added, which takes it to 1.037... We can stop here. As soon as this second sum exceeds 1, the Kelly recipe will never include the horse that took the sum above one, or any later horses.

Now for each k where $A(k) < 1$, work out the ratio

$$B(k) = \frac{1 - P(k)}{1 - A(k)},$$

and compare it with the value of $p \times (\alpha + 1)$ for the horse whose label is k. Provided that $p \times (\alpha + 1)$ exceeds $B(k)$, we shall include all the horses $1, 2, \ldots, k$ in our bet. In fact, suppose K is the largest value of k for which this condition holds. The Kelly strategy places particular bets on the horses $1, 2, \ldots, K$.

In the example, since $A(1) < 1$, we find $B(1) = (1 - 0.2)/(1 - 2/11) = 0.977..$, and for horse C we know that $p \times (\alpha + 1) = 1.1$, which is bigger than $B(1)$. Our bet will include horse C. Next $B(2) = (1 - 0.6)/(1 - 0.56277) = 0.91485$; and for horse A, $p \times (\alpha + 1)$ is 1.05, which exceeds $B(2)$. We add horse A to our bets. Also $B(3) = (1 - 0.9)/(1 - 0.87046) = 0.772$; and since for horse B, the value of $p \times (\alpha + 1)$, 0.975, exceeds this, we shall include the third horse B in our best bet, *even though a bet on B alone is not favourable.* We cannot go further, since $A(4)$ is bigger than 1. The Kelly strategy is to bet on horses C, A, and B. Here K, the number of horses we spread our bets on, is three.

The final step is to decide how much to bet on each horse. We bet in total a fraction $1 - B(K)$ of our fortune; on horse k, bet the fraction $p(k) - B(K)/(\alpha (k) + 1)$, for $k = 1, 2, \ldots, K$. In the example, we should bet $0.2 - 0.772 \times 2/11 = 6\%$ on C; and $0.4 - 0.772 \times 8/21 = 10.6\%$ on A; and $0.3 - 0.772 \times 4/13 = 6.25\%$ on B.

Chapter 11 takes this same example, modified so that horse B is now offered at odds of 2:1 only, with the odds for A and C remaining fixed, and all five probabilities of winning being the same. Carry through the Kelly recipe for this changed position, and check that you have got it right by comparing your answer with the one given there. The main point is that now the third horse B no longer forms part of your optimal bet.

Solutions to test yourself quizzes

4.1 The chance of n Heads and n Tails is close to $1/\sqrt{n\pi}$.

    (i)   50 tosses means $n = 25$, so the chance is close to $\sqrt{25\pi} \approx 11\%$.

    (ii)   250 tosses means $n = 125$. Answer is about 5%.

    (iii) 1000 tosses means $n = 500$. Chance is about 2.5%.

4.2 If he chooses HHHH, you should select THHH. Unless the first four tosses are H, you must win, so your chance is 15/16. (This is like the argument in the text with HHH.) If he selects TTHH, the hint in the text suggests you make TTH the *last* three in your choice, so either HTTH or TTTH ought to be best. The other hint, to avoid palindromes, leads to TTTH. (Using methods not described in the book, your winning chance is then 2/3; with HTTH, your winning chance is 7/12.)

4.3 For 2, 4, 6, 8, and 10 tosses, the ratios of the corresponding chances in tables 4.1 and 4.4 are 1, 3, 5, 7, and 9.

4.4 Granted this pattern goes on for ever, the ratios after 50, 250, and 1000 tosses will be 49, 249, and 999. So, using the answers to Q4.1, the chance that the *first* return to equality is after 50 tosses is 11%/49, about 0.002; for 250 and 1000 tosses, the corresponding chances are about 0.0002 and 0.000025—this is about 1 in 40 000.

4.5 Zero lead changes is more likely than any other particular number. Since H and T can only be equal after an even number of tosses, the answer for 1001 tosses is the same as for 1000. But, at toss 1000, the most likely time of last equality is at the two extremes, i.e. after zero tosses, or after 1000 tosses. After 1001 tosses, the guesses '0' and '1000' are equally good, and better than any other guess.

4.6 We know the chance of exactly x lead changes decreases as x increases. Since each of B, C, and D contain exactly ten values, then no matter how many tosses are made, D is more likely than C, and C is more likely than B.

With N tosses, the guidelines for the breakeven point are just a little higher than $\sqrt{N/3}$. When $N = 1001$, this break even point is about 12—we are about equally likely to have more than 12, or fewer than 12 lead changes. Since A asks for at least 30, this strongly suggests A is very unlikely. The order looks like D, C, B, A (and it is). But when N is 10 001, this break even point is above 30, so the chance of A exceeds 50%. The order must be A, D, C, B.

6.1 The row minima are 4, 7, and 3, so the maximin is 7. The column maxima are 12, 10, and 15, so the minimax is 10. There is no saddle point. Replacing '15' by '6' means that the maximum in the third column is now 7, so the minimax is 7, and (II, C) is a saddle point.

6.2 With Table 6.6, the column differences are $(6, -2)$, which lead to $(2, 6)$, i.e. Roy selects H with probability $2/8 = 1/4$. The row differences are $(4, -4)$, which lead to $(4, 4)$, so Colin uses H and T equally often.

6.3 There are no dominated strategies at all. But if the '10' were 9 or less, II would dominate III.

6.4 With the payoffs

	B	C
I	4	15
II	9	7

the column differences are $(11, -2)$ so Rows uses I and II with probabilities $(2/13, 11/13)$. The row differences are $(5, -8)$ so Columns uses B and C with probabilities $(8/13, 5/13)$. When Rows uses this strategy, the mean payoff is 107/13 whatever Columns does, so Rows should pay $107/13 \approx 8.23$ units each game to make it fair.

6.5 A die has only six faces, so to get ten equally likely alternatives, we must use it at least twice. Here is one method. Roll the die until you do *not* get a Six, and note the answer, A. Then roll it once more. If this last roll is 1, 2, or 3, the answer is A; if the last roll is 4, 5, or 6, the answer is $A + 5$.

6.6

	0	1	2	3	4
0	1	1	1	-1	-1
1	1	1	-1	-1	-1
2	1	-1	-1	-1	1
3	-1	-1	-1	1	1
4	-1	-1	1	1	1

For Roy, 0 dominates 1 and 4 dominates 3. So cross out the rows labelled 1 and 3, to get

	0	1	2	3	4
0	1	1	1	−1	−1
2	1	−1	−1	−1	1
4	−1	−1	1	1	1

Colin prefers negative entries, so for him 1 dominates 0, and 3 dominates 4. Cross out the columns labelled 0 and 4.

	1	2	3
0	1	1	−1
2	−1	−1	−1
4	−1	1	1

Roy can now eliminate 2, dominated by either of the other two strategies, which then means Colin can also throw out 2, leaving simply

	1	3
0	1	−1
4	−1	1

which is equivalent to the fair game of Table 6.1. Roy chooses from 0 and 4 at random, Colin selects from 1 and 3 at random.

6.7 Colin will show 2 coins every time, and always win. Roy would be a fool to play.

6.8 Consider a Hawk–Hawk contest. Each has a 50% chance of winning so the average gain to either is +6, but each suffers 8 units of damage: net result is –2. For Dove–Dove, the net result is +3 (+6 for the 50% winning chance, –3 for the time wasting.) Thus

	H	D
H	−2	12
D	0	3

With a proportion p of H, the average H payoff is $-2p + 12(1 - p) = 12 - 14p$, and the average payoff to D is $3 - 3p$. These are equal when $p = 9/11$. A population that plays Hawk with probability 9/11 is at an ESS.

6.9 If the prize is worth $2x$, the H–H payoff is $x - 8$, the D–D payoff is $x - 3$. In H–D, H gets $2x$ and D gets 0. If ever x exceeds 8, H does better than D at any population composition, so Doves are eliminated. The prize can only be worth having when x *is* positive; suppose it is less than 8. Then we always have an ESS with both H and D, and the proportion of H will be $(x + 3)/11$. This does make sense—the smaller x, the fewer Hawks.

10.1 Bold play is: whenever you have £50 or less, bet the entire amount; with more than £50, bet just enough to hit £100 if you win. Starting with £20, write $p = 73/148$, the chance you win on a bet on Red. The chance you win twice, and hit £80, is p^2; you might then Win, or Lose then Win, or Lose twice which takes you back where you started. Write x = chance you achieve your goal. The above shows that

$$x = p^2(p + (1 - p)p + (1 - p)^2X)$$

which leads to

$$x(1 - p^2(1 - p)^2) = p^3(2 - p)$$

and thus to $x = 0.19286..$—not far short of $x = 0.2$, which is your chance in a perfectly fair game!

If your bet of £16 at 5–1 wins, you have exactly £100. The chance it wins is 6/37. If it loses (chance 31/37), your bet of £4 on zero has chance 1/37 of taking you to £144. Altogether, the chance is 0.1848.

Yes, you could do just better. You could place just £3 on zero if your first bet loses, which still takes you to your target when it wins, and leaves you £1 if it loses. With that last £1, try an 11–1 shot on a row of three numbers, followed by an 8–1 bet on a block of four if successful. That adds another $(3/37) \times (4/37) = 0.0088$, taking your chance up to 0.1936, *just* better than bets on Red!

10.2 Your £100 will last for 100 spins, even if you are unlucky every time. Since you lose, on average, £(1/37) each spin, your average loss after 90 spins is £2.43. To be ahead after 89 spins, you need five or more wins, but to be ahead after 91 spins, you need six or more. Use Trick 1 to work out the chance of at most four wins, and at most five wins, respectively. Write x = 2/37 and y = 35/37 as your chances of win or lose on one spin. Use the binomial distribution (Appendix II).

89 spins: Probability of four or fewer wins is

$y^{89} + 89y^{88}x + 3916y^{87}x^2 + 113564y^{86}x^3 + 2441626y^{85}x^4$ which evaluates to 0.47019. So the chance you are ahead is about 53%.

91 spins: Probability of five or fewer wins is

$y^{91} + 91y^{90}x + 4095y^{89}x^2 + 121485y^{88}x^3 + 2672670y^{87}x^4 +$

$46504458y^{86}x^5$ which is 0.63063... The chance you are ahead is down to about 37%.

10.3 The chance of reaching home is just $20/100 = 20\%$. The average number of steps is $20 \times 80 = 1600$. For his brother, the chance of safety is the same (40/200 is still 20%), but the average number of steps is $40 \times 160 = 6400$, four times as many.

10.4 A. You should raise with a Pair. The dealer has a qualifying hand, but your Pair is better than his. You win three units in total (one on your ante, and two at evens on your raise).

    B. You lose three units. You will raise, but his hand is better.

    C. You raise, and win the showdown. A Full House pays at 8–1, so your total profit is 17 units.

Literature cited

Appleton, D. R. (1995). May the best man win? *The Statistician*, **44**, 529–38.

Aldous, D. and Diaconis, P. (1986). Shuffling cards and stopping times. *American Mathematical Monthly*, **93**, 333–48.

Bagley, D. (1969). *The Spoilers*. Collins.

Bass, T. A. (1990). *The Newtonian Casino*. Longman.

Bayer, D. and Diaconis, P. (1992). Trailing the dovetail shuffle to its lair. *Annals of Applied Probability*, **2**, 294–313.

Benford, F. (1938). The law of anomalous numbers. *Proceedings of the American Philosophical Society*, **78**, 551–72.

Blackwell, D., Deuel, P., and Freedman, D. (1964). The last return to equilibrium in a coin-tossing game. *Annals of Statistics*, **35**, 1344.

Clutton-Brock, T. H. and Albon, S. D. (1979). The roaring of red deer and the evolution of honest advertisement. *Behaviour*, **69**, 145–70.

Clutton-Brock, T. H., Albon, S. D., Gibson, R. M., and Guinness, F. E. (1979). The logical stag: adaptive aspects of fighting in red deer (*Cervus elaphus L.*). *Animal Behaviour*, **27**, 211–25.

Coe, P. R. and Butterworth, W. (1995). Optimal stopping in 'The Showcase Showdown'. *The American Statistician*, **49**, 271–5.

Cover, T. A. (1989). Do longer games favour the stronger player? *The American Statistician*, **43**, 277–8.

David, F. N. (1962). *Games, Gods and Gambling*. Griffin.

Dawkins, R. (1976). *The Selfish Gene*. Oxford University Press.

Downton, F. and Holder, R. L. (1972). Banker's games and the Gaming Act 1968. *Journal of the Royal Statistical Society, Series A*, **135**, 336–56.

Dostoyevsky, F. M. (1866). *The Gambler*.

Epstein, R. A. (1967). *The Theory of Gambling and Statistical Logic*. Academic Press, New York and London.

Feller, W. (1950, 1957, 1968). *An Introduction to Probability Theory and its Applications*, vol. 1. All three editions, John Wiley and Sons.

Gabriel, K. R. and Newman, J. (1962). A Markov chain model for daily rainfall occurrence in Tel Aviv. *Quarterly Journal of the Royal Meteorological Society*, **88**, 90–5.

George, S. L. (1973). Optimal strategy in tennis. *Applied Statistics*, **22**, 97–104.

Gould, S. J. (1996). *Dinosaur in a Haystack*. Jonathan Cape.

Greene, G. (1955). *Loser Takes All*. Heinemann.

Griffin, P. A. and Gwynn, J. M. Jr (1998). An analysis of Caribbean stud poker. In *Finding the Edge: Mathematical and Quantitative Analysis of Gambling* (ed. Judy A. Cornelius).

Haigh, J. (1995). Random Hex, and Blockbusters. *The Mathematical Scientist*, **20**, 91–102.

Haigh, J. (1996). More on n-point, win-by-k games. *Journal of Applied Probability* **33**, 382–7.

Holder, R. L. and Downton, F. (1972). Casino pontoon. *Journal of the Royal Statistical Society, Series A*, **135**, 221–41.

Kelly, J. L. Jr. (1956). A new interpretation of information rate. *Bell System Technical Journal*, **35**, 917–26.

Kemeny, J. G. and Snell, J. L. (1957). Game-theoretic solution of baccarat. *American Mathematical Monthly*, **64**, 465–9.

Lowe, D. A. (ed. and trans.) (1989, 1990). *Fyodor Dostoyevsky. Complete letters*, vols. 2 and 3. Ardis, Ann Arbor.

Markov, A. A. (1913). An example of statistical analysis of the text of *Eugene Onegin*, illustrating the association of trials into a chain. *Bulletin de l'Académie Imperiale des Sciences de St Petersburg*, **7**, series 6,153–62.

Maynard Smith, J. (1982). *Evolution and the Theory of Games*. Cambridge University Press.

Millman, M. H. (1983). A statistical analysis of casino blackjack. *American Mathematical Monthly*, **90**, 431–6.

Morris, C. (1977). The most important points in tennis. In *Optimal Strategies in Sports* (ed. S. P. Lahany and R. E. Machol). North Holland Elsevier.

Newcombe, S. (1881). Note on the frequency of use of the different digits in natural numbers. *American Journal of Mathematics*, **4**, 39–40.

Nigrini, M. J. (1992). The detection of income evasion through an analysis of digital distributions. Ph.D. thesis, Dept of Accounting, University of Cincinnati.

Packel, E. (1981). *The Mathematics of Games and Gambling*. The Mathematical Association of America, New Mathematics Library.

Raimi, R. A. (1976). The first digit problem. *American Mathematical Monthly*, **83** 521–6.

Ridder, G., Cramer, J. S. and Hopstaken, P. (1994). Estimating the effect of a red card in soccer. *Journal of the American Statistical Association*, **89**, 1124–7.

Riedwyl, H. (1990). *Zahlenlotto*. Paul Haupt, Bern.

Simon, J. (1997). An analysis of the distribution of combinations chosen by National Lottery players. Working paper, European University Institute, Florence.

Thorp, E. O. (1962). *Beat the Dealer*. Blaisdell, New York.

Varian, H. R. (1972). Benford's law. *The American Statistician*, 26, 65.

Waugh, E. (1931). *Remote People*.

Weaver, W. (1962). *Lady Luck*. Heinemann.

Williams, J. D. (1966). *The Compleat Strategyst*. McGraw-Hill, New York, San Francisco.

Index